THE

POETRY OF SCIENCE,

OR

STUDIES OF THE

PHYSICAL PHENOMENA OF NATURE.

BY

ROBERT HUNT,

AUTHOR OF 'PANTHEA,' 'RESEARCHES ON LIGHT,' ETC.

How charming is Divine Philosophy!
Not harsh and crabbed as dull fools suppose,
But musical as is Apollo's lute,
And a perpetual feast of nectar'd sweets,
Where no crude surfeit reigns.

MILTON.

S. H. White

FIRST AMERICAN FROM THE SECOND LONDON EDITION.

BOSTON:
GOULD, KENDALL, AND LINCOLN,
59 WASHINGTON STREET.
1850.

From Shakspeare to Plato — from the philosophic poet to the poetic philosopher — the transition is easy, and the road is crowded with illustrations of our present subject.

Hast thou ever raised thy mind to the consideration of EXISTENCE, in and by itself, as the mere act of existing?

Hast thou ever said to thyself thoughtfully IT IS! heedless, in that moment, whether it were a man before thee, or a flower, or a grain of sand; — without reference, in short, to this or that particular mode or form of existence? If thou hast, indeed, attained to this, thou wilt have felt the presence of a mystery, which must have fixed thy spirit in awe and wonder.

COLERIDGE.

PREFACE.

A Second Edition of this work being demanded within a twelvemonth of the publication of the first, convinces the author that he has not erred in believing it possible to interest a numerous class of readers by a popular examination of the deductions of philosophy and the truths of inductive science; — that he was not mistaken in believing the generalizations from mechanical experiments to be capable of assuming a poetic aspect.

Desiring to present every discovery of Science with correctness, increased attention has been paid to the present edition; and it has been cleared of some trifling inaccuracies — regretted by no one more than the author — which had crept into the first.

To several friends of eminence in their respective branches of chemistry, geology, and natural history, the author has to express his warmest thanks. To them is due the emendation of all that was obscure; and the aid they have afforded in the revision of those passages which relate to their own familiar studies must give to this Edition an essential value.

To keep pace with the progress of science, each new discovery, connecting itself with the physical laws of nature, has been included;—although not numerous, they are important.

It is, therefore, hoped that the author's attempt to render the "POETRY OF SCIENCE" a correct text-book, at the same time he has striven to divest science of its harder features, will not have been unsuccessful.

R. H.

November, 1849.

PREFACE TO THE FIRST EDITION.

An attempt has been made, in this volume, to link together those scientific facts which bear directly and visibly upon Natural Phenomena, and to show that they have a value superior to their mere economic applications, in their power of exalting the mind to the contemplation of the Universe.

In pursuing this object, where the links of the inductive chain were imperfect, a consideration of the merits of conflicting theories has been hazarded; and — probably as a consequence of imperfect knowledge — the views entertained by those whose scientific acquirements no one esteems more highly than myself, have, in a few instances, been departed from.

The authority for each statement is given at the conclusion of the volume; and an experimental examination has been made of all the instances adduced in exemplification of particular views.

For the purpose of exhibiting the great facts of Science in their most attractive aspects, the imagination has been occasionally taxed, but it has never been allowed to interfere with the stern reality of Truth; and throughout the following pages it is hoped there will be discovered — however illogical to some, certain speculations may appear — evidences of a constant endeavour to think honestly, and to give to every discovery its full value in the generalizations attempted.

R. H.

November, 1848.

CONTENTS.

CHAPTER I.

GENERAL CONDITIONS OF MATTER.

CHAPTER II.

MOTION.

CHAPTER III.

GRAVITATION.

CHAPTER IV.

MOLECULAR FORCES.

CHAPTER V.

CRYSTALLOGENIC FORCES.

CHAPTER VI.

HEAT — SOLAR AND TERRESTRIAL.

CHAPTER VII.

LIGHT.

CHAPTER VIII.

ACTINISM — CHEMICAL RADIATIONS.

CHAPTER IX.

ELECTRICITY.

CHAPTER X.

MAGNETISM.

CHAPTER XI.

CHEMICAL FORCES.

CHAPTER XII.

CHEMICAL PHENOMENA.

CHAPTER XIII.

TIME — GEOLOGICAL PHENOMENA.

CHAPTER XIV.

PHENOMENA OF VEGETABLE LIFE.

CHAPTER XV.

PHENOMENA OF ANIMAL LIFE.

CHAPTER XVI.

GENERAL CONCLUSIONS.

INTRODUCTION.

The True is the Beautiful. Whenever this becomes evident to our senses, its influences are of a soul-elevating character. The beautiful, whether it is perceived in the external forms of matter, associated in the harmonies of light and colour, appreciated in the modulations of sweet sounds, or mingled with those influences which are, as the inner life of creation, appealing to the soul through the vesture which covers all things, is the natural theme of the poet, and the chosen study of the philosopher.

But, it will be asked, where is the relation between the stern labours of science and the ethereal system which constitutes poetry? The fumes of the laboratory, its alkalies and acids, the mechanical appliances of the observatory, its specula and its lenses, do not appear fitted for a place in the painted bowers of the Muses. But, from the labours of the chemist in his cell, — from the multitudinous observations of the astronomer on his tower, — spring truths which the philosopher employs to interpret nature's mysteries, and which give to the soul of the poet those realities to which he aspires in his high imaginings.

Science solicits from the material world, by the persuasion of inductive search, a development of its elementary principles, and of the laws which these obey. Philosophy strives to apply the discovered facts to the great phenomena of being, — to deduce large generalities from the fragmentary discoveries of severe induction, — and thus to ascend from matter and its properties up to those impulses which stir the whole, floating, as it were, on the confines of sense, and indicating, though dimly, those superior powers which, more nearly related to infinity, mysteriously manifest themselves in the phenomena of mind. Poetry seizes the facts of the one and the theories of the other; unites them by a pleasing thought, which appeals for truth to the most unthinking soul, and leads the reflective intellect to higher and higher exercises; it connects common phenonema with exalted ideas; and, applying its holiest powers, it invests the human mind with the sovereign strength of the True.

Truth is the soul of the poet's thought; — truth is the reward of the philosopher's toil; and their works, bearing this official stamp, live among men through all time. Science at present rejoices in her ministry to the requirements of advancing civilization, and is content to receive the reward given to applications which increase the comforts of life, or add to its luxuries. Every improvement in the arts or manufactures has a tendency to elevate the race who are benefited thereby. But because science is useful in the working days of our week, it is not to be neglected on our Sabbath, — when, resting from our labours, it becomes agreeable to contemplate the few truths permitted to our knowledge, and thus enter into communion, as closely as is allowed to finite beings, with those influences which involve and interpenetrate the earth, giving to all things Life, Beauty, and Divinity.

The human mind naturally delights in the discovery of truth; and even when perverted by the constant operations of prevailing errors, a glimpse of the Real comes upon it like the smile of daylight to the sorrowing captive of some dark prison. The Psychean labours to try man's soul, and exalt it, are the search for truth beneath the mysteries which surround creation, — to gather amaranths, shining with the hues of heaven, from plains upon which hang, dark and heavy, the mists of earth. The poet may pay the debt of nature, — the philosopher may return to the bosom of our common mother, — even their names fade in the passage of time, like planets blotted out of heaven; — but the truths they have revealed to man burn on for ever with unextinguishable brightness. Truth cannot die; it passes from mind to mind, imparting light in its progress, and constantly renewing its own brightness during its diffusion. The True is the Beautiful; and the truths revealed to the mind render us capable of perceiving new beauties on the earth. The gladness of truth is like the ringing voice of a joyous child, and the most remote recesses echo with the cheerful sound. To be forever true is the Science of Poetry, — the revelation of truth is the Poetry of Science.

Man, a creation endued with mighty faculties, but a mystery to himself, stands in the midst of a wonderful world, and an infinite variety of phenomena arise around him, in strange form and magical disposition, like the phantasma of a restless night.

The solid rock obeys a power which brings its congeries of atoms into a thousand shapes, each one geometrically perfect. Its vegetable covering, in obedience to some external excitation, developes itself in a curious diversity of forms, from the exquisitely graceful to the singularly grotesque, and exhibits properties still more varied and opposed. The animal organism quickened by higher impulses, — powers working within, and modifying the influence of the external forces, — presents, from the Monad to the Mammoth, and through every phase of being up to Man, a yet more wonderful series of combinations, and features still more strangely contrasted.

Lifting our searching gaze into the measureless space beyond our earth, we find planet bound to planet, and system chained to system, all impelled by a universal force to roll in regularity and order around a common centre. The pendulations of the remotest star are communicated through the unseen bond; and our rocking world obeys the mysterious impulse throughout all those forces which regulate the inorganic combinations of this earth, and unto which its organic creation is irresistibly compelled to bow.

The glorious sun by day, and the moon and stars in the silence and the mystery of night, are felt to influence all material nature, holding the great Earth bound in a many-stranded cord which cannot be broken. The tidal flow of the vast ocean, with its variety of animal and vegetable life, — the atmosphere, bright with light, obscured by the storm-cloud, spanned by the rainbow, or rent with the explosions of electric fire, — attest to the might of these elementary bonds.

These are but a few of the great phenomena which play their part around this globe of ours, exciting men to wonder, or shaking them with terror.

The mind of man, in its progress towards its higher destiny, is tasked with the physical earth as a problem, which, within the limits of a life, it must struggle to solve. The intellectual spirit is capable of embracing all finite things. Man is gifted with powers for studying the entire circle of visible creation; and he is equal, under proper training, to the task of examining much of the secret machinery which stirs the whole.

In dim outshadowing, earth's first poets, from the loveliness of external nature, evoked beautiful spiritualizations. To them, the shady forests teemed with aërial beings,—the gushing springs rejoiced in fantastic sprites,—the leaping cataracts gleamed with translucent shades,—the cavernous hills were the abodes of genii,—and the earth-girdling ocean was guarded by mysterious forms. Such were the creations of the far-searching mind in its early consciousness of the existence of unseen powers. The philosopher picked out his way through the dark and labyrinthine path, between effects and causes, and slowly approaching towards the light, he gathered semblances of the great Reality, like a mirage, beautiful and truthful, although still but a cloud-reflection of the vast Unseen.

It is thus that the human mind advances from the Ideal to the Real, and that the poet becomes the philosopher, and the philosopher rises into the poet; but at the same time as we progress from fable to fact, much of the soul-sentiment which made the romantic holy, and gave a noble tone to every aspiration, is too frequently merged in a cheerless philosophy which clings to the earth, and reduces the mind to a mechanical condition, delighting in the accumulation of facts, regardless of the great laws by which these are regulated, and the harmony of all Telluric combinations secured. In science, we find the elements of the most exalted poetry; and in the mysterious workings of the physical forces, we discover connections with the illimitable world of thought,—in which mighty minds delight to try their powers,—as strangely complicated, and as marvellously ordered, as in the psychological phenomena which have, almost exclusively, been the objects of their studies.

In the aspect of visible nature, with its wonderful diversity of form and its charm of colour, we find the Beautiful; and in the operations of these principles, which are ever active in producing and maintaining the existing conditions of matter, we discover the Sublime.

The form and colour of a flower may excite our admiration; but when we come to examine all the phenomena which combine to produce that piece of symmetry and that lovely hue,—to learn the physiological arrangement of its structural parts,—the chemical actions by which its woody fibre and its juices are produced,—and to investigate those laws by which is regulated the power to throw back the white sunbeam from its surface in coloured rays,—our admiration passes to the higher feeling of deep astonishment at the perfection of the processes, and of reverence for their great Designer. There are, indeed, "tongues in trees;" but science alone can interpret their mysterious whispers, and in this consists its poetry.

To rest content with the bare enunciation of a truth, is to perform but one half of a task. As each atom of matter is involved in an atmosphere of properties and powers, which unites it to every mass of the universe, so each truth, however common it may be, is surrounded by impulses which, being awakened, pass from soul to soul like musical undulations, and which will be repeated through the echoes of space, and prolonged for all eternity.

The poetry which springs from the contemplation of the agencies which are actively employed in producing the transformations of matter, and which is founded upon the truths developed by the aids of science, should be in no respect inferior to that which has been inspired by the beauty of the individual forms of matter, and the pleasing character of their combinations.

The imaginative view of man and his world — the creations of the romantic mind — have been, and ever will be, dwelt on with a soul-absorbing passion. The mystery of our being, and the mystery of our ceasing to be, acting upon intelligences which are for ever striving to comprehend the enigma of themselves, leads by a natural process to a love for the Ideal. The discovery of those truths which advance the human mind towards that point of knowledge to which all its secret longings tend, should excite a higher feeling than any mere creation of the fancy, how beautiful soever it may be. The phenomena of Reality are more startling than the phantoms of the Ideal. Truth is stranger than fiction. Surely many of the discoveries of science which relate to the combinations of matter, and exhibit results which we could not by any previous efforts of reasoning dare to reckon on, results which show the admirable balance of the forces of nature, and the might of their uncontrolled power, exhibit to our senses subjects for contemplation truly poetic in their character.

We tremble when the thunder-cloud bursts in fury above

our heads. The poet seizes on the terrors of the storm to add to the interest of his verse. Fancy paints a storm-king, and the genius of romance clothes his demons in lightnings, and they are heralded by thunders. These wild imaginings have been the delight of mankind; there is subject for wonder in them: but is there anything less wonderful in the well-authenticated fact, that the dewdrop which glistens on the flower, that the tear which trembles on the eyelid, holds locked in its transparent cells an amount of electric fire equal to that which is discharged during a storm from a thunder-cloud?

In these studies of the effects which are continually presenting themselves to the observing eye, and of the phenomena of causes, as far as they are revealed by Science in its search of the physical earth, it will be shown that beneath the beautiful vesture of the external world there exists, like its quickening soul, a pervading power, assuming the most varied aspects, giving to the whole its life and loveliness, and linking every portion of this material mass in a common bond with some great universal principle beyond our knowledge. Whether by the improvement of the powers of the human mind, man will ever be enabled to embrace within his knowledge the laws which regulate these remote principles, we are not sufficiently advanced in intelligence to determine. But if admitted even to a clear perception of the theoretical Power which we regard as regulating the known forces, we must still see an unknown agency beyond us, which can only be referred to the Creator's will.

THE

POETRY OF SCIENCE.

CHAPTER I.

GENERAL CONDITIONS OF MATTER.

Its varied Characters, and constant change of external Form — The Grain of Dust, its Properties and Powers — Resulting Combinations in inorganic Masses and in organized Creations — Our knowledge of Matter — Theory of Ultimate Atoms — The Physical Forces acting on the Composition of Masses — The certainty of the exercise of subtile principles, which are beyond the reach of experimental Science.

THE Physical Earth presents to us, in every form of organic and inorganic matter, an infinite variety of phenomena. If we select specimens of rocks, either crystalline or stratified, — of metals in any of their various combinations with oxygen, sulphur, and other bodies, — of gems glistening with light and glowing with colour, — or if we examine the varied forms and hues of the vegetable world, and the more mysterious animal creations, we must inevitably come to a conclusion long since proclaimed, and admit that dust they are, and that to dust must they return. Whatever permanency may be given to matter, it is certain that its form is ever in a state of change. The surface of the "Eternal Hills" is worn away by the soft rains which fall to fertilize; and from their wrecks, borne by the waters to the ocean, new continents are forming. The mutations of the old earth may be read upon her rocks and mountains, and these records of former changes tell us the infallible truth, that as the present passes into the future, so will the form of nature undergo an important alteration. The same forces which lifted the Andes and the Himalayas are still at work, and from the particles of matter carried from the present lands by the rivers into the sea, where they subside in stratified masses, there will, in the great future, be raised new worlds, upon which the work of life may go forward, and over which will be spread a vast Intelligence.

If we regard the conditions of the beautiful and varied organic covering of the Earth, the certainty, the constancy, of change is ever before us. Vegetable life passes into the animal form, and both perish to feed the future plant. Man, moving to-day the monarch of a mighty people, in a few years passes back to *his primitive* clod, and that combination of elementary atoms, which is dignified with the circle of sovereignty and the robe of purple, after a period may be sought for in the herbage of the fields, and in the humble flowers of the valley.

We have, then, this certain truth, — all things visible around us are aggregations of atoms. From particles of dust, which under the microscope could scarcely be distinguished one from the other, are all the varied forms of nature created. This grain of dust, this particle of sand, has strange properties and powers. Science has discovered some, but still more truths are hidden within this irregular molecule of matter which we now survey, than even philosophy dares dream of. How strangely it obeys the impulses of heat — mysterious are the influences of light upon it — electricity wonderfully excites it — and still more curious is the manner in which it obeys the magic of chemical force. These are phenomena which we have seen; we know them, and we can reproduce them at our pleasure. We have advanced a little way into the secrets of nature, and from the spot we have gained, we look forward with a vision somewhat brightened by our task; but we discover so much to be yet unknown, that we learn another truth, — our vast ignorance of many things relating to this grain of dust.

It gathers around it other particles; they cling together, and each acting upon every other one, and all of them arranging themselves around the little centre, according to some law, a beautiful crystal results, the geometric perfection of its form being a source of admiration.

It exerts some other powers, and atom cohering to atom, obeying the influences of many external radiant forces, undergoes inexplicable changes, and the same dust which we find forming the diamond, aggregates into the lordly tree, — blends to produce the graceful, scented, and richly painted flower, — and combines to yield the luxury of fruit.

It quickens with yet undiscovered energies; it moves with life: dust and vital force combine; blood and bone, nerve and muscle, result from the combination. Forces, which we cannot by the utmost refinements of our philosophy detect, direct the whole, and from the same dust which formed the rock and grew in the tree, is produced a living and a breathing thing, capable of receiving a Divine illumination, of bearing in its new state the gladness and the glory of a Soul.

These considerations lead us to reflect on the amount of our knowledge. We are led to ask ourselves, what do we know? We know that the world with all its variety is composed of certain material atoms, which, although presented to us in a great variety of forms, do not in all probability differ very essentially from each other.

We know that those atoms obey certain conditions which appear to be dependent upon the influences of motion, gravitation, heat, light, electricity, and chemical force. These powers are only known to us by their effects; we only detect their action by their operations upon matter; and although we regard the several phenomena which we have discovered, as the manifestations of different principles, it is possible they may be but modifications of some one universal power, of which these are but a few of its modes of action.

In examining, therefore, the truths which science has revealed to us, it is advantageous, for the purpose of fixing the mind to the subject, that we assume certain conditions as undeniably true. These may be stated in a few sentences, and without wasting a thought upon those metaphysical subtleties which have from time to time perplexed science, and served to impede the progress of truth, we proceed, then, to examine our knowledge of the phenomena which constantly occur around us.

Every form, whether inorganic or organic, which we can discover within the limits of human search, is composed of hard impenetrable atoms, which are capable of assuming, under the influence of certain physical forces, conditions essential to the composition of that body of which it forms a part. ([1]) The known forces, active in producing these conditions, are modes of motion; gravitation and aggregation, heat, light; and associated with these, actinism or chemical radiation; electricity, under all its conditions, whether static or dynamic; and chemical affinity, regarded as the result of a separate elementary principle.

These forces must be considered as powers capable of acting in perfect independence of each other. They are possibly modifications of one principle; but this view being an hypothesis, which, as yet, is only supported by loose analogies, cannot, without danger, be received in any explanation which attempts to deal only with the truths of science.

In addition to the known physical forces, we cannot examine the varied phenomena of nature, without feeling that there must be other and most active principles of a higher order than any detected by science, to which belong the important operations of vitality, whether manifested in the plant or the animal. In treating of these, although speculation cannot be entirely avoided, it will be employed only so far as it gives any assistance in linking phenomena together.

We have to deal with the active agencies which give form and feature to nature — which regulate the harmony and beauty and vigor of life — and upon which depend those grand changes in the conditions of matter, which must convince us that death is but the commencement of a new state of Being.

CHAPTER II.

MOTION.

Are the Physical Forces modes of Motion ? — Motion defined — Philosophical views of Motion, and the Principles to which it has been referred — Motions of the Earth and of the Solar System — Influence of the proper Motions of the Earth on the Conditions of Matter — Theory of the Conversion of Motion into Heat, &c. — The Physical Forces regarded as principles independent of Motion, although the Cause and often apparently the Effects of it.

MANY of the most eminent thinkers of the present time are disposed to regard all the active principles of nature as "modes of motion,"— to look upon light, heat, electricity, and even vital force, as phenomena resulting from "change of place" among the particles of matter; this change, disturbance, or motion, being dependent upon some undefined mover.([2])

The habit of leaving purely inductive examination for the delusive charms of hypothesis — of viewing the material world as a metaphysical bundle of essential properties, and nothing more — has led some of our most eminent philosophers to struggle with the task of proving, that all the wonderful manifestations of the great physical powers of the universe are mere modifications of motion.([3])

The views of metaphysicians regarding motion involve many subtle considerations which need not at present detain us. We can only consider motion as a change of place, which matter cannot effect of itself, no change of place being possible without a mover; and, consequently, motion cannot be a *property* of matter in the strict sense in which that term should be accepted.([4])

Motion depends upon certain disturbing and directing forces acting upon all matter; and, consequently, as every mode of action is determined by some excitement external to the body moved, motion cannot, philosophically, be regarded otherwise than as a peculiar affection of matter under determinable conditions. "We find," says Sir Isaac Newton, "but little motion in the world, except what plainly flows from either the active principles of nature, or from the command of the willer."(5)

Plato, Aristotle, and the Pythagoreans supposed that throughout all nature an active principle was diffused, upon which depended all the properties exhibited by matter. This is the same as the "plastic nature" of Cudworth,(6) the "intellectual and artificial fire" of Bishop Berkeley;(7) and to these all modes of motion were referred. Sir Isaac Newton also regards the material universe and its phenomena as dependent upon "*active principles* — for instance, the cause of gravity — whereby the planets and comets preserve their motions in their orbits, and all bodies acquire a degree of motion in falling; and the cause of fomentation — whereby the heart and blood of animals preserve a perpetual warmth and motion — the inner parts of the earth are kept constantly warmed — many bodies burn and shine — and the sun himself burns and shines, and with his light warms and cheers all things."

The earth turns on its axis at the rate of more than 1,000 miles an hour, and passes around the sun with the speed of upwards of 68,000 miles in the same time.(8) The earth and the other planets of our system move in curves around a common centre; therefore their motion cannot have been originally communicated merely by the impressed force of projection. Two forces, at least, must have operated, one making the planets tend directly to the centre, and the other impelling them to fly off at a tangent to the curve described. Here we have a system of spheres, held by some power to a great central mass, around which they revolve with a fearful velocity. Nor is this all; the Solar System itself, bound by the same mystic chain to an undiscovered centre, moves towards a point in space at the rate of 33,550,000 geographical miles, whilst our earth performs one revolution around the sun. ([9])

In addition to these great rotations, the earth is subjected to other motions, as the precession of the equinoxes and the nutation of its axis. Rocking regularly upon a point round which it rapidly revolves, whilst it progresses onward in its orbit, like some huge top in tremulous gyration upon the deck of a vast aërial ship gliding rapidly through space, is the earth performing its part in the great law of motion.

The rapidity of these impulses, supposing the powers of the physical forces were for a moment suspended, would be sufficient to scatter the mass of our planet over space as a mere star-dust.

Limiting, as much as possible, the view which opens upon the mind as we contemplate the adjustments by which this great machine, our system, is preserved in all its order and beauty, let us forget the great movement of the whole through space, and endeavour to consider the effect of those motions which are directly related to the earth, as a member of one small group of worlds.

We cannot for a moment doubt, although we have not any experimental proof of the fact, that the proper motions of the earth materially influence the conditions of the matter of which it is formed. Every pair of atoms are, like a balance, delicately suspended, under the constant struggle which arises from the tendency to fly asunder, induced by one order of forces, and the efforts of others to chain them together. The spring is brought to the highest state of tension—one tremor more, and it would be destroyed.

We cannot, by any comparison with the labors of the most

skilful human artisan, convey an idea of the exquisite perfection of planetary mechanics, even so far as they have been discovered by the labors of science; and we must admit that our insight into the vast machinery has been very limited.

All we know, is the fact that this planet moves in a certain order, and at a fixed rate, and that the speed is of itself sufficient to rend the hardest rocks; yet the delicate down which rests so lightly upon the flower is undisturbed. It is, therefore, evident that matter in all its forms is endued with powers, by which mass is bound to mass, and atom to atom; these powers are not the results of any of the motions which we have examined, but, acting in antagonism to them, they sustain our globe in its present form.

Are there other motions to which these powers can be referred? We know of none. That absolute rest may not exist among the particles of matter is probable. Electrical action, chemical power, crystalline aggregation, the expansive force of heat, and many other known agencies are in constant operation to prevent it. It must, however, be remembered, that each and every atom constituting a mass may be so suspended between the balanced forces, that it may be regarded as relatively at rest.

Theory imagines Motion as producing Force — a body is moved, and its mere mechanical change of place is regarded as generating heat; and hence the refinements of modern science have advanced to the conclusion that motion and heat are convertible. Admitting that the material atoms of which this world is formed are never in a state of quiescence, yet we cannot suppose any gross ponderable particle as capable of moving itself; but once set in motion, it may become the secondary cause of motion in other particles.([10]) The difficulties of the case would appear to have been as follows: Are heat, light, electricity, &c. material bodies? If they are material bodies — and heat, for example, is the cause of motion — must not the calorific matter move itself — or if it be not self-moving, by what is it moved? If heat is material, and the primary cause of motion, then matter must have an innate power of moving; it can convert itself into active force, or be at once a cause and an effect, which can scarcely be regarded as a logical deduction.

We move a particle of matter, and heat is manifested; the force being continued, light, electricity, and chemical action result; all, as appears from a limited view of the phenomena, arising out of the mechanical force applied to the particle first moved.([11]) This mechanical force, it must be remembered, is external to the body moved, and is, in all probability, set up by the movement of a muscle, acted upon by nerves, under the influence of a will.

The series of phenomena we have supposed to arise, admit of an explanation free of the hypothesis of motion, and we avoid the dangerous ground of metaphysical speculation, and the subtleties of that logic which rests upon the immateriality of creation. This explanation, it is freely admitted, is incomplete: we cannot distinctly correllate each feature of the phenomena, combine link to link, and thus form a perfect chain; but it is sufficiently clear to exhibit what we do know, and leave the unknown free for unbiassed investigation.

Each particle, each atom of that which conveys to our senses the only ideas we have of natural objects — ponderable matter — is involved in, or interpenetrated by, those principles which we call heat or electricity, and probably many others which are unknown to us; and although these principles or powers are, according to some law, bound in statical equilibrium to inert matter, they are freely developed by an external excitement, and the disturbance of any one of them, upsetting the equilibrium, leaves the other powers equally free to be brought under the cognizance of human sense.

When we come to an examination of the influences exerted by these powers upon the physical earth, the position, that they must be regarded as the causes of motion rather than the effects of it, will be further considered. At present, it is only necessary to state thus generally the views we entertain of the conditions of matter in connection with the imponderable forces and mechanical powers. The conversion, as it has been called, of motion into heat, in the experiments of Count Rumford and Mr. Joule,([12]) are only evidences that a certain uniformity exists between the mechanical force applied, and the amount of heat liberated. It does not appear that we have any proof of the conversion of motion into physical power.

It is necessary, to anything like a satisfactory contemplation of the wonderful properties of matter, and of the forces regulating the forms, of the entire creation, that we should be content with regarding the elementary bodies which chemistry instructs us form our globe, as tangible, ponderable atoms, having specific and distinguishing properties. That we should, as far as it is possible for finite minds to do so, endeavour to conceive the powers or forces — gravitation, molecular attraction, electricity, heat, light, and the principle which determines all chemical phenomena — as manifestations of agencies which hold a place between the most subtile form of matter and the spiritual state, which reveals itself dimly in psychological phenomena, and arrives at its sublimity in the God of the universe.

CHAPTER III.

GRAVITATION.

The forms of Matter — Shape of the Earth — Probability of the Mass forming this Planet having existed in a Nebulous State — Zodiacal Light — Comets — Volatilization of Solid Matter by Artificial means — The principle of Gravitation — Its Influence through Space and within the smallest Limits — Gravitating powers of the Planets — Density of the Earth — Certainty of Newton's Law of the Inverse Square — Discovery of Neptune — State of a Body independent of Gravitation — Experiment explaining Saturn's Ring, &c. — General inference.

Let us suppose the earth, consisting of three conditions of matter — the solid, the fluid, and the aëriform — to be set free from that power by which it is retained in its present form of a spheroid flattened at the poles, but still subject to the influences of its diurnal and annual rotations. Agreeably to the law which regulates the conditions of all bodies moving at high velocities, the consequence of such a state of things would be, that our planet would instantly spread itself over an enormous area. The waters and the solid masses of this globe would, in all probability, present themselves amidst the other phenomena of space in a highly attenuated state, revolving in an orbit around the sun, or as a band of nebulous matter, which might sometimes be rendered sensible to sight by condensation in the form of flights of shooting stars. ([13])

This may be illustrated by experiment. If upon a rapidly revolving disc we place a ball of dust, it will be almost immediately spread out, and its particles will arrange themselves in a series of regular curves, varying with the velocity of the motion. In addition to the disintegration which would arise from the tendency of the atoms to fly from the centre, the motion, in space, of the planetary mass would naturally occasion a trailing out, and the only degree of uniformity which this orb could, under these imaginary conditions, possibly present, would be derived from the combined effects of dissimilar motions.

Amid the remoter stars, some remarkable cloud-like appearances are discovered. These nebulæ, presenting to the eye of the observer only a gleaming light, as from some phosphorescent vapour, were long regarded as indications of such a condition as that which we have just been considering. Astronomers saw, in those mysterious nebulæ, a confirmation of their views, which regarded all the orbs of the firmament as having once been thin sheets of vapour, which had gradually, from irregular bodies traversing space, been slowly condensed about a centre, and brought within the limits of aggregating agencies, until, after the lapse of ages, they became sphered stars, moving in harmony amid the bright host of heaven.(14) Geologists seized on those views with eagerness, as confirming theoretical conclusions deduced from an examination of the structure of the earth itself, and explained by them the gradual accretion of atoms into crystalline or stratified rocks.

The researches of modern astronomers, aided by the magnificent instruments of Lord Rosse,(15) have, however, shown that many of the most remarkable nebulæ are only clusters of stars, so remote from us, that the light from them appears blended into one diffusive sheet or luminous film. There are, however, the Magellanic clouds, and other singular patches of light, exhibiting changes which can only be explained on the theory of their slow condensation. There is no evidence to disprove the position that world-formation may still be going on; that a slow and gradual aggregation of particles, under the influence of laws with which we are acquainted, may be constantly in progress, to end, eventually, in the formation of a sphere.

May we not regard the zodiacal light as the remains of a solar luminiferous atmosphere, which once embraced the entire system of which it is the centre?([16]) Will not the strange changes which have been *seen to take place* in cometary bodies, even whilst they were passing near the earth, as the division of Biela's comet and the ultimate formation of a second nucleus from the detached portion, strongly tend to support the probability of the idea that attenuated matter has, in the progress of time, been condensed into solid masses, and that nebulous clouds must still exist in every state of tenuity in the regions of infinite space,([17]) which, in the mysterious processes of world-formation, will, eventually, become stars, and reflect across the blue immensity of heaven, in brightness, that light which is the necessary agent of organization and all manifestations of beauty.

The inferences drawn from a careful study of the condition of our own globe, is in favor of the assumption of the existence of nebulous matter. By the processes of art and manufacture, by the operation of those powers on which organization and life depend, solid matter is constantly poured off in such a state that it cannot be detected, *as matter*, by any of the human senses. Yet a thousand results, daily and hourly accumulating as truths around us, prove that the solid metals, the gross earths, and the constituents of animal and vegetable life, all pass away invisible to us, and become "thin air." We know that, floating around us, these volatilized bodies exist in some form or other, and numerous experiments in chemistry are calculated to convince us, that the most attenuated air is capable, with a slight change of circumstances, of being converted into the condition of solid masses. Hydrogen gas, the lightest, the most etherial of the chemical elements, dissolves iron and zinc, arsenic, sulphur, and carbon; and from the transparent combinations thus formed, we can with facility separate those ponderous bodies. Such substances must exist in our own atmosphere; why not in the regions of space? Whether this planet ever floated a mass of nebulous matter, only known by its dim and filmy light, or comet-like rushed through space with eccentric orbit, are questions which can only receive the reply of speculative minds. Whether the earth and the other members of the Solar System were ever parts of a Central Sun,(18) and thrown from it by some mighty convulsion, though now revolving with all the other masses around that orb, chained in their circuits by some infinite power, is beyond the utmost refinements of science to discover. This hypothesis is, however, in its sublime conception, worthy of the master-mind that gave it birth.

All we know is, that our earth is a sphere, which, by the effects of its motion, is somewhat enlarged at the equator and depressed at the poles; — that it maintains its regular course around the sun, in virtue of a force which, acting constantly, would eventually draw it into the body of the sun itself; but that this force is opposed by the momentum of the revolving mass; — that the same force acting from the centre of the earth itself, and from the centre of every particle of its substance, resolves the whole into a globular form.

The principle of Gravitation(19) is that property of matter by which particle is attracted by particle, and mass by mass, the less towards the greater. What this may be, we scarcely dare to speculate. In the vast area of its action, which opens before the eye of the mind, we see a power spanning all space, and linking together every one of those myriads of worlds which spangle the robe of the Infinite, and we are compelled to pause. Is this principle of gravitation a property of matter, or is it a power higher than the more tangible forces, is the question which presses on the mind. If we regard it as a subtile principle pervading all space, we compel ourselves to look beyond it for another power yet more refined; and we cannot halt until, ascending from the limitable to the illimitable, we resolve gravitation and its governing influence to the centre of all power — the will of the eternal Creator.

Science has developed the grand truth, that it is by the exercise of this all-pervading influence that the earth is retained in its orbit — that the crystal globe of dew which glistens on the leaf is bound together — that the débris which float upon the lake accumulates into one mass — that the sea exhibits the phenomena of the tides — and the aërial ocean its barometric changes. In all things this force is active, and throughout nature it is ever present. Our knowledge of the laws which it obeys, enables us to conclude that the sun and distant planets are consolidated masses like this earth. We find that they have gravitating power, and by comparing this influence with that exerted by the earth, we are enabled to weigh the mass of one planet against another. In the balance of the astronomer, it is as easy to poise the remote star, as it is for the engineer to calculate the weight of the iron tunnel of the Menai Straits, or any other mechanical structure. Thus throughout the universe the balance of gravitating force is unerringly sustained. If one of the most remote of those gems of light, which flicker at midnight in the dark distance of the starry vault, was, by any power, removed from its place, the disturbance of these delicately balanced mysteries would be felt through all the created systems of worlds.

From the peculiarity of the laws which this power called gravity obeys, it has been inferred that it acts from centres of force; it is proved that its power diminishes in the inverse ratio of the square of the distance, and that the gravitating power of every material body is in the direct proportion of its mass. In astronomical calculations we have first to learn the mass of our earth. Experiment informs us that the density of our hardest rock is not above 2.8 ; but from the enormous pressure to which matter must be subjected, at great depths from the surface, the weight of the superincumbent mass constantly increasing, it is quite certain that the earth's density must be far more than this. Maskelyne suspended a plummet over Schehallion,(20) and Cavendish, with exceedingly delicate apparatus, observed the attraction of masses of known weight and size upon each other; and applying the powers of arithmetical calculation, and the data obtained from the small experiments to the larger phenomena, the first determined the earth's mean density to be 4.71, whilst the latter made it 5.48, and the more recent refined investigations of Baily have determined it to be 5.67.(21)

From data thus obtained by severe inductive experiments and mathematical analyses, the astronomer, by observing the deviations of a distant star from a true path, is enabled to determine the influence of those stellar bodies near which it passes, and, hence, to calculate the relative magnitude of each. The accuracy of the law is in this way put to the severest test, and the precision of astronomical prediction is the strongest proof of its universality and truth.

Rolling onward its lonely way, in the far immensity of our system, the planet Uranus was discovered by the elder Herschel, —so great its distance that its diminished light could scarcely be detected by the most powerful telescopes; but since its discovery its path has been carefully watched, and some irregularities noticed. Most of these disturbances were referable to known causes; but a little trembling observed when the planet was in one portion of its vast orbit was unexplained. Convinced of the certainty of Newton's law, these deviations were referred to the gravitating influence of a mass unknown to us; and by the investigations of Adams in England,(22) and Le Verrier in France,(23) the place of an hypothetical planet was determined; and as a grand confirmation of the great law, and to the glory of those two far-searching minds, who do honor to their respective countries and their age, the hypothesis became a fact, in the discovery of the planet Neptune in the place determined by rigorous calculation. Astronomy affords other examples of the sublime truth of the law of gravitation, than which science can afford no more elevated poetry.

So completely is all nature locked in the bonds of this infinite power, that it is no poetic exaggeration to declare, that the blow which rends any earthly mass is conveyed by successive impulses to every one of the myriads of orbs, which are even too remote for the reach of telescopic vision.

An illustrative experiment must close our consideration of this remarkable principle. We well know that a body in a fluid state would, if suspended above the earth, it being at the same time free to take any form, naturally assume that of a flattened spheroid, from the action of the mass of the earth upon it; whereas the force of cohesive attraction acting equally from all sides of a centre, would necessarily produce a perfect sphere. The best method of showing that this would be the case, is as follows: —

Alcohol and water are to be mixed together until the fluid is of the same specific gravity as olive oil. If, when this is effected, we drop globules of the oil into the mixed fluid, it will be seen that they take an orbicular form; — and, of course, in this experiment the power of the earth's gravitating influence is neutralized. The same drops of oil under any other conditions would be flattened. Simple as this illustration is, it tells much of the wondrous secret of those beautifully balanced forces of cohesion and of gravitation; and from the prosaic fact we rise to a great philosophical truth. Our experiment may lead us yet farther in exemplification of known phenomena. If we pass an iron wire through one of those floating spheres of oil, and make it revolve rapidly, imitating the motion of a planet on its axis, the oil spreads out, and we have the spheroidal form of our earth. Increase the rapidity of this rotation, and when a certain rate is obtained the oil divides, and a ring, connected by the finest possible film with the central globe, revolves around it. (24) Here we have a minute representation of the ring of Saturn. This is a suggestive experiment, the repetition of which, by reflective minds, cannot fail to lead to important deductions. The phenomena of cohesion, of motion, and gravitation, are all involved; and we produce results resembling, in a striking manner, the conditions which prevail in the planetary spaces, under the influence of the same powers.

From the centre of our earth to the utmost extremity of the universe—from the infinitely small to the immensely vast—gravitation exerts its force. It is met on all sides by physical powers acting in antagonism to it, but like a ruling spirit it restrains them all.

The smallest dust which floats upon the wind
Bears this strong impress of the eternal mind.
In mystery round it, subtile forces roll;
And gravitation binds and guides the whole.
In every sand, before the tempest hurled,
Lie locked the powers which regulate a world,
And from each atom human thought may rise
With might to pierce the mysteries of the skies,—
To try each force which rules the mighty plan,
Of moving planets, or of breathing man;
And from the secret wonders of each sod,
Evoke the truths, and learn the power of God.

CHAPTER IV.

MOLECULAR FORCES.

Conditions of Matter — Variety of organized Forms — Inorganic Forms — All matter reducible to the most simple conditions — Transmutation, a natural operation — Chemical Elementary Principles — Divisibility of Matter — Atoms — Molecules — Particles — Molecular Force includes several Agencies — Instanced in the Action of Heat on Bodies — All Bodies porous — Solution — Mixture — Combination — Centres of Force — Different states of Matter (Allotropic Conditions) — Theories of Franklin, Æpinus, and Coulomb — Electrical and Magnetic Agencies — Ancient Notions — Cohesive Attraction, &c.

In contemplating the works of nature, we cannot but regard, with feelings of religious admiration, the infinite variety of forms under which matter is presented to our senses. On every hand the utmost diversity is exhibited; through all things we trace the most perfect order; and over all is diffused the charm of beauty. It is the uneducated or depraved alone who find deformities in the creations by which we are surrounded.

The three conditions of matter are — the solid, the fluid, and the aëriform; and these belong equally to the organic and the inorganic world.

In organic nature we have an almost infinite variety of animal form, presenting developments widely different from each other, yet in every case suited to the circumstances required by the position which the creature occupies in the scale of being. Through the entire series, from the Polype to the higher order of animals, even to man, we find a uniformity in the progress towards perfection, and a continuity in the series, which betrays the great secret, that the mystery of life is the same in all,—a pervading spiritual essence associated with matter, and modifying it by the master-mechanism of an Infinite mind.

In the vegetable clothing of the surface of the earth, which fits it for the abode of man and animals — from the confervæ of a stagnant pool, or the lichen of the wind-beaten rock, to the lordly oak or towering palm — a singularly beautiful chain of being, and of gradual elevation in the scale of organization, presents itself to the contemplative mind.

In the inorganic world, where the great phenomena of life are wanting, we have constantly exhibited the working of powers of a strangely complicated kind. The symmetrical arrangement of crystals — the diversified characters of mineral formations — the systematic aggregation of particles to form masses possessing properties of a peculiar and striking nature — all prove that agencies, which science, with all its refinements, has not yet detected, are unceasingly at work. Heat, electricity, chemical power — whatever that may be — and the forces of cohesion, are known to be involved in the production of the forms we see; but contemplation soon leads to the conviction that these powers are subordinate to others which we know not of. We know only the things belonging to the surface of our planet, and these but superficially. The geologist traces rock-formations succeeding each other (from the primary strata holding no traces of organized forms, through the Paleozoic series, in which, step by step, the history of animal life is recorded,) to the more recent formations, teeming with relics, which, though allied to some animal types still quickened with life, are generally such as have passed away. The naturalist searches the earth, the waters, and the air, for their living things; and the diversity of form, the variety of condition, and the perfection of organization which he discovers as belonging to this our epoch — differing from, indeed bearing but a slight relation to, those which mark the earth's mutations — exhibit, in a most striking view, the endless variety of characters which matter can assume.

We are so accustomed to all these phenomena of matter, that it is with some difficulty we can bend ourselves to the study of the more simple conditions in which it exists.

The solid crust of this telluric sphere — the waters and the atmosphere — the diversified fabrics of the vegetable kingdom — and the still more complicated structures of men and animals — are, altogether, but the aggregation of minute particles in accordance with certain fixed laws. By mechanical means all kinds of matter may be reduced to powder, the fine particles of which would not appear very different from each other; but each atom has been impressed with properties peculiar to itself, which man has no power to change.

To nature alone belongs the mysterious property of transmutation. The enthusiastic alchemist, by the agency of physical forces, dissipates a metal in vapour; but it remains a metal, and the same metal still. By the Hermetic art, he breaks up the combination of masses; but he cannot alter the principles of any one of the elements which formed the mass upon which his skill is tried.

Every atom is invested with properties peculiar to all of its class; and each one possesses powers, to which in mute obedience it is compelled, by which these properties are modified, and the character of matter varied. What are those properties? Do we know anything of those powers?

The earth, so far as we are acquainted with it, is composed of about sixty principles, which we call elementary. These are the most simple states to which we can reduce matter, and from them all the forms of creation yet examined by the chemist are produced. These elementary principles are, some of them, permanently gaseous under the ordinary temperature, and others exist as solid masses; the difference between the two conditions being regulated, as it appears, by the opposing forces of heat and cohesive attraction.

Matter has been regarded by some as infinitely divisable; but the known conditions of chemical combinations lead to the conclusion that there are limits beyond which matter cannot be divided.([25]) The theory of atoms having determinate characters, and possessing symmetric forms, certainly has the advantage of presenting to the human mind a starting-point — a sort of standing ground, from which it can direct the survey of cosmical phenomena. The metaphysical hypothesis, which resolves all matter into properties, and refers all things to ideas, leaves the mind in a state of uncertainty and bewilderment.

Adopting the views of Dumas, with some modifications,([26]) it will be found more satisfactory to regard the *ultimate atoms* of matter as points beyond the reach of our examination; which, according to a law determined by the influences of the so called imponderable forces, unite to form *molecules*. Again, these molecules combine to form the *particles* of the mass which we may regard as the limit of mechanical division.

The particles of solid bodies are solid, those of fluids fluid, and those of gaseous bodies are themselves aëriform; but it does not follow that the molecules of any body should be necessarily solid, fluid, or aëriform, from the circumstance of their having formed the particles of a body in one of these states.

As this planet — a molecule in space — is formed of aggregated atoms, and enveloped by its own physical agencies — and as it is involved in the infinitely extending influences of other planetary molecules, and thus forms part of a system — so the molecules of any mass are grouped into a system or particle, which possesses the great characteristic features of the whole.

In an aëriform body the particles are in a state of extreme tenuity, the molecules being themselves, by the influence of some repulsive force, just on the verge where cohesion exerts its decaying power. In fluid bodies, the attenuation of the particles is less — the particles and also the molecules are nearer together. Whereas, in the solid body, the forces of cohesion are most strongly exerted, and all the molecular conditions brought more powerfully into action.

Under the term molecular force, we include several agencies which are not alike, but which are all-powerful in producing the general characteristics of bodies. These require a somewhat close examination. All the particles of even a solid mass are at a distance from each other, and are free to move. By heat we can increase the length and thickness of a bar of iron, or any other metal. Fluids and gases in like manner obey the dispersive influence of caloric. From these and other analogous results we learn that all bodies have a greater or less degree of porosity. The distance at which the particles of fluid bodies are maintained, is strikingly proved by the fact, that hydrated salts dissolved in water occupy no more space than that which is equal to the water contained in the crystalline body. All the solid matter of the salt must, in these cases, go to fill up the interstitial spaces of the fluid.([27])

The conditions which regulate the solubility of bodies, and the power of solution, regarded either as a mechanical or a chemical process, are very obscure. We might be led to suppose, that those bodies possessing the largest amount of unoccupied space were capable of holding the greatest quantity of soluble matter dissolved. This is not the case as a general rule.

The peculiar manner in which hydrogen gas appears to dissolve solid substances, as iron, potassium, sodium, sulphur, phosphorus, selenium, and arsenic, is explained by regarding the result as one of a chemical character, and only a manifestation of the powers of affinity over the forms of bodies. In like manner the solution of salt in water, or the mixture of alcohol in that fluid, may be viewed as a chemical phenomenon, although usually considered as simple cases of solution or mixture. Alterations of temperature and other physical changes take place in all. If two masses of metal, tin and copper for example, are melted and combined, the united mass will not equal the bulk of the two masses. If a pint measure of oil of vitriol and an equal quantity of water are mixed together, the combined fluids will not fill a two pint measure. ([28])

In these instances a large quantity of heat is rendered sensible, as if it had been squeezed out, by the force with which the particles united, from interstices which it would appear were filled with what we may be allowed to call an atmosphere of heat. Hence we conclude that, amongst the influences determining the molecular constitution of a body, heat performs an important part. All these facts go to prove that the atoms which form the compound body, whatever may be its character, are disposed of as so many centres of force, which act by influences of a peculiar character upon each other. That these influences are dependent upon known physical forces is certain; but the laws by which the powers of the ultimate atom are altered remain still unknown.

In the great operations of nature, changes are produced which we cannot understand, and variations of condition do certainly occur, which may be regarded as instances of transmutation.

Amongst others, we may adduce the different states in which we know carbon to exist. We have the diamond with its beautiful light-refracting property, its hardness, and high specific gravity, capable of being converted into graphite and coke. ([29]) Charcoal, graphite, and the diamond are totally unlike each other, yet we know they are each composed of the same atoms. Charcoal is a black irregular substance, light, and readily inflammable; graphite is crystallizable; but the forms of its crystals cannot be referred to those of the diamond, and it burns with difficulty. The diamond occurs in the most regular and beautifully transparent forms; and it can be burned only at the highest artificial temperatures. We are, however, convinced by experiment that the brilliant and transparent gem is made up of the same atoms as those which go to form the dull black mass of charcoal. What is the mystery of this? We know not. These peculiar conditions have been the subjects of anxious study; but science has not yet let in a ray of light upon the mystery. That a different state — it has been called an *allotropic* condition — is often induced in the same class of atoms is certain; and hence the variety of the resulting compounds. To continue our illustrations with carbon — may not its combinations, in uniform proportions with oxygen and hydrogen,(30) owe their differences to some allotropic change in the ultimate atoms of this element?

We know that silicium — the metallic base of flint — is capable of assuming two or more different states; and sulphur, selenium, phosphorus, and arsenic are susceptible of these remarkable changes. Copper, iron, tin, and manganese are known to exist in at least two states, and many of the rarer metals exhibit the same peculiarity.(31) Hence, may we not infer that some of those substances, which we now term elementary, are but dissimilar conditions of the same element? The remarkable resemblance between many of those bodies strengthens the speculation. Iridium and platinum, iron and nickel, chlorine, bromine, iodine, and probably fluorine, are good examples of these similarities, although these bodies are all distinguished by physical and chemical differences.

The light-refracting gem, which glistens on the neck of beauty, and is valued for its pure transparency, differs only from the rude lump of coke in its molecular arrangement. Chemistry teaches us that we may, without producing any disarrangement of affinities, but merely by setting up molecular disturbance, effect decided changes, as is strikingly shown in the colour of iodide of mercury, under the influence of heat; — and by a slight change, merely molecular, produced by caloric, iron may be made to resemble platinum in its chemical relations.(32) On studying this question we certainly find good reason for supposing that bodies, resembling each other in most of their properties, are the result of different conditions which have been impressed upon the ultimate atoms, similar to those discovered in the substances we have named. This hypothesis appears to be more in accordance with the great principles which we must conceive guided the labours of an Infinite Mind, than that which supposes a vast number of individual creations. It will be seen in the sequel that light, heat, electricity, and chemical action have the power of producing yet more striking changes in the forms of bodies; and it is probable that, according to the operations of these agents, either combined or separate, acting over different spaces of time, and under varying circumstances, in relation to the molecular forces, all those allotropic states may be produced. Hence bodies may still be discovered, which, from the imperfections of science, resisting our means of decomposition, must, for a time, be regarded as new elements.

The experiments of Faraday prove that all matter is in certain polar conditions, having apparently the powers of mutual attraction and repulsion.(33) Are the molecular forces to be referred to any of those powers? Are they not probably the result of some ultimate principle of which these properties are but the modified manifestations?

Franklin supposed the minute atoms of bodies to be surrounded with a fluid or ether, which they condensed upon their surfaces with great force — and we have experiments showing that this is the case—([34]) whilst he regarded the atoms of the ether itself as mutually repellent, thus establishing an equilibrium of forces. Æpinus reduced the hypothesis of Franklin to a mathematical theory; and Coulomb *proved* that the force with which the repulsion of the ethereal atoms and the attraction of the material molecules are produced, is, like universal attraction, to whatever power that may be due, regulated by the law of the inverse ratio of the square of the distance. These views are found, upon minute examination, to hold true to the phenomena with which inductive science has made us acquainted; and the striking manner in which, when submitted to the rigorous investigations of geometers, they agree with known conditions of electricity, appears certainly to favor the opinion that this power may be materially connected with these molecular arrangements.

Many of the phenomena which are connected with the magnetic influences, also bear in a remarkable manner upon this inquiry. But, without the necessary proof of direct experimental evidence, it were as unphilosophical to refer the binding together of the molecules of matter to the agency of electricity, in any of its modes, as it would be to adopt the theory of the hooked atoms of Epicurus, or the astrological dream of the sympathies of matter.([35])

Science, however, enables us to infer with safety that the mechanism which regulates the constitution of a cube of marble, or a granite mountain, is of the same order as that which determines the earth's place in the solar system, and the situation of the solar system in the immensity of space.

In fine, cohesion, or the attraction of aggregation, is a power employed in binding particle to particle. To cohesion, we find we have caloric opposed as a repellent force; and the mysterious operations of those electrical phenomena, generally referred to as polar forces, are constantly, it is certain, interfering with the powers of cohesion. In addition, we have seen that in nature there exists an agency which is capable of changing the constitution of the ultimate atoms, and of thus giving variety to each resulting mass. What this power may be, our science cannot tell; but our reason leads us, with firm conviction, to the belief that it is a principle which is, beyond all others in its subtile influences, manifesting to us the divine power of the omniscient Creator.

The molecular forces involve a consideration of all the known physical powers, the study of which, in their operations on matter, will engage our attention. But it is pleasant to learn, as we advance, step by step in our examination of the phenomena of creation, that we may study the grand in what externally appears the simple, and learn, in the mysteries of a particle, the high truths which science has to tell of a planet.

It may appear that we regard the forces of gravitation and cohesion as identical. Many phenomena, which we are enabled to reach by the refinements of inductive inquiry, certainly present to us a striking similarity in the laws which regulate the operations of these powers; but it must be remembered that their identity is not established. To quote the words of Young: "The whole of our inquiries respecting the intimate nature of forces of any kind must be considered merely as speculative amusements, which are of no further utility, than as they make our views more general, and assist our experimental investigations."([36])

CHAPTER V.

CRYSTALLOGENIC FORCES.

Crystallization and Molecular Force distinguished — Experimental Proof — Polarity of Particles forming a Crystal — Difference between Organic and Inorganic Forms — Decomposition of Crystals in Nature — Substitution of Particles in Crystals — Pseudomorphism — Crystalline Form not dependent on Chemical Nature — Isomorphism — Dimorphism — Theories of Crystallogenic attraction — Influence of Electricity and Magnetism — Phenomena during Crystallization — Can a change of Form take place in Primitive Atoms? — Illustrative example of Crystallization.

"CRYSTALLIZATION is a peculiar and most admirable work of nature's geometry, worthy of being studied with all the power of genius, and the whole energy of the mind, not on account of the delight which always attends the knowledge of wonders, but because of its vast importance in revealing to us the secrets of nature; for here she does, as it were, betray herself, and, laying aside all disguise, permits us to behold, not merely the results of her operations, but the very processes themselves." Such is the language of an Italian philosopher, Gulielmini, and it is the striking peculiarity of beholding the process of the formation of the regular geometric figures of crystals, the gradual accretion of particle to particle, which induces us to separate crystallization from mere molecular aggregation. Without doubt the formation of a crystal and the production of an amorphous block are due to powers which bear a close resemblance in many points; but they present remarkable differences in others.

Let us take some simple case in illustration. In quiet water we have very finely divided matter suspended, and matter in a state of solution. The first is slowly precipitated, and in process of time consolidates into a hard mass at the bottom, presenting no particular character, unless it has been placed in peculiar physical conditions; when, as in nature, we have a regular bedding which is intersected by lines of lamination or of cleavage, which we are, from experiment, enabled to refer to the influence of current electricity. The second — the matter in solution — is also slowly deposited; but it is accumulated upon nuclei which possess some peculiar disposing powers, and every particle is united by some particular face, and an angular figure of the most perfect character results. Many pleasing experiments would appear to show that electricity has much to do in the process of crystallization; but it is evident that it must be under some peculiarly modified conditions that this power is exerted, if, indeed, it has any direct action.

The same substances always crystallize in the same forms, unless the conditions of the crystallizing body are altered. It has been supposed that each particle of a crystalline mass has certain points or poles which possess definite properties, and that cohesion takes place only along lines which have some relation to the attracting or repelling powers of these poles. We shall have, eventually, to consider results which appear to prove that magnetism is universal in its influence, and that this polarity of the particles of matter may be referred to it.

Be the cause of crystallization what it may, it presents to us a near approach in inorganic nature to some of the peculiar functions of organized creation. In the one, we have the gradual accretion of parts and the formation of members due to peculiar powers of assimilation, each individual preserving all its distinguishing features; and in the other, we have a regular order of cohesion occurring under the influence of a power which draws like to like, and arranges the whole into a form of beauty.

We must, however, remember that a striking difference exists between the productions of the mineral and the other kingdoms of nature. Animals and vegetables arrive at maturity by successive developments, and increase by the assimilation of substances, having the power of producing the most important chemical changes upon such matter as comes with the range of their influence; but minerals are equally perfect in the earliest stages of their formation, and increase only by the accretion of particles without their undergoing any change.

The animal and vegetable tribes cease to continue the functions of life: death ensues, and a complete disorganization takes place; but this is not the case in the mineral world: the crystal being the result of a constantly acting force is not necessarily liable to decomposition. Nevertheless, we find in nature that crystals, after arriving at what may be regarded as, in some sort, their maturity, are, owing to a change of the conditions under which they were formed, gradually decomposed. In our mines we find skeletons of crystals, and within the cavity thus formed, others of a different constitution and figure find nuclei, and the conditions required for their development. Again, to give a striking instance, the felspar crystals of the granitic formations are liable to decomposition in a somewhat peculiar manner. In decomposing, these crystals leave moulds of their own peculiar forms, and it not unfrequently happens, in the stanniferous districts of Cornwall, that oxide of tin gradually fills these moulds, and we procure this metallic mineral in the form of the earthy one. Again, we have the curious instances of bodies crystallizing in a false form under change of circumstances. We find, for example, Pseudomorphism, as this peculiar class of phenomena is named, occurring by the removal of the constituent atoms of one body, while another set, which naturally assumes a different form, takes their place, yet still preserving the original shape. It often happens that copper pyrites will, in this manner, exhibit the angles of an ordinary variety of crystallized carbonate of iron. These curious changes may be familiarized by supposing a beautiful form of gold, which some skilful mechanic takes to pieces, particle by particle, so skillfully substituting a grain of brass for every one of gold removed, that the loss of the precious metal cannot be detected by any mere examination of its form.

Crystalline form is not strictly dependent upon the chemical nature of the parts forming the crystal. The same number of atoms, arranged in the same way, produce the same form. Substances much unlike each other will assume the same crystalline arrangement. Magnesia, lime, oxide of cadmium, the protoxides of iron, nickel, and cobalt combined with the same acid, present similarly formed bodies. These Isomorphic([37]) peculiarities are so common that the discoverer of the phenomena, Mitscherlich, announces as a law, "that the chemical elements of which all bodies consist are susceptible of being classified in distinct groups."

We also find compounds which have two distinct systems of crystallization. This property, Dimorphism, is very strikingly shown in carbonate of lime, which occurs in rhombohedrons, in calc spar, and in rhombic prisms in arragonite.

Crystals are found of the most microscopic character, and of an exceedingly large size. A crystal of quartz at Milan is three feet and a quarter long, and five feet and a half in circumference, and its weight is 870 pounds. Beryls have been found in New Hampshire measuring four feet in length.([38])

In the dark recesses of the earth, where the influences which produce organization and life cease to act, a creative spirit still pursues its never-ending task of giving form to matter.

The science of crystallogeny,([39]) embracing the theoretical and practical question of the causes producing these geometric forms, has in various ways attempted to explain the laws according to which molecules arrange themselves on molecules in perfect order, giving rise to a rigidly correct system of architecture. But it cannot be said that any theory yet propounded is sufficiently exact to embrace the whole of the known phenomena, and the questions, — What is crystallogenic attraction, and what is the physical nature of the ultimate particles of matter, — are still open for the inquiries of that genius which delights in wrestling with the secrets of nature.

The great Epicurus speculated on the "plastic nature" of atoms, and attributed to this *nature* the power they possess of arranging themselves into symmetric forms. Modern philosophers satisfy themselves with attraction, and, reasoning from analogy, imagine that each atom has a polar system.

Electricity, and light, and heat, exert remarkable powers, and both accelerate and retard crystallization; and we have recently obtained evidence which appears to prove that some form of magnetism has an active influence in determining the natural forms of crystals. Electricity appears to quicken the process of crystalline aggregation — to collect more readily together those atoms which seek to combine — to bring them all within the limits of that influence by which their symmetrical forms are determined; and strong evidence is now afforded, in support of the theory of magnetic polarity, by the refined investigations of Faraday and Plücker, which prove that magnetism has a *directing* influence upon crystalline bodies in direct dependence upon the crystalline or optic axes of the body.([40])

It has been found that crystals of sulphate of iron, slowly forming from a solution which has been placed within the range of powerful magnetic force, dispose themselves along certain magnetic curves; whereas the Arbor Dianæ, or silver tree, forming under the same circumstances, takes a position nearly at right angles to these curves. Certain groups of crystals have been found in nature, which appear to show, by their positions, that terrestrial magnetism has been active in producing the phenomena they exhibit.([41])

During rapid crystallization, some salts — as the sulphate of soda, boracic acid, and arsenious acid crystallizing in muriatic acid — exhibit decided indications of electrical excitement. Light is given out in flashes, and we have evidence that crystals exhibit a tendency to move towards the light. Professor Plücker has ascertained that certain crystals — in particular the cyanite — "point very well to the north, by the magnetic power of the earth only. It is a true compass needle; and, more than that, you may obtain its declination." We must remember that this crystal, the cyanite, is a compound of silica and alumina only. This is the amount of experimental evidence which science has afforded in explanation of the conditions under which nature pursues her wondrous work of crystal formation. We see just sufficient of the operation to be convinced that the luminous star which shines in the brightness of heaven, and the cavern-secreted gem, are equally the result of forces which are known to us in only a few of their modifications.

Every substance, when placed under circumstances which allow of the free movement of its molecules, has a tendency to crystallize. All the metals may, by slowly cooling from the melting state, be exhibited with a crystalline structure. Of the metallic and earthy minerals, nature furnishes us with an almost infinite variety of crystals, and by a reduction of temperature, yet more simple bodies assume the most symmetric forms. Water, in the conditions of ice and snow, is a familiar and beautiful example; and, by such extreme degrees of cold as are artificially produced, many of the gases exhibit a tendency to a crystalline condition.

May not the solid elementary atoms be susceptible of change of form under different influences? May not the different states under which the same bodies are found — as, for example, silica, carbon, and iron — be due entirely to a change in the form of the primitive atom?

Admitting the probability of this, we then easily see that the central molecule, formed of an aggregation of such atoms, uniting by particular faces, would present a determinate form; and that the resulting crystal, a mass of such molecules, cohering according to a given law, at certain angles, would present such geometric figures as we find in nature, or produce in our laboratories, when we avail ourselves of processes which nature has taught us.

If we take a particle of marble, and place it in a large quantity of water acidulated with sulphuric acid, it dissolves, and a new compound results. The marble disappears — the eye cannot detect it by form or colour: the acid also has been disguised — the taste discovers nothing sour in the fluid. We have, in combination with the water, the lime and sulphuric acid; but that combination appears to the eye in no respect different from the water itself. It is colourless and perfectly transparent, although it holds a mass of solid matter which previously would not allow of the passage of a ray of light. Let us expose this fluid to such circumstances that the water will slowly evaporate, and we shall find forming in it, after a time, microscopic particles of solid, light-refracting matter. These particles gradually increase in size, and we may watch their growth until eventually we have a symmetric figure, beautifully shaped, the primary form of which is a right rhomboidal prism. Thus in nature, by the action, in all probability, of vegetable matter on the sulphates held in solution by the water of the great rivers and the ocean — aided by our oxidizing atmosphere — sulphuric acid is formed to do its work upon the limestone formations, and from this combination would result the well-known gypsum, or plaster of Paris, which ordinarily exists as an amorphous mass, but is often found in a crystalline form. (42)

This is a very perfect illustration of the wonderful process we have been considering, and in which, simple though it appears to be, we have set to work a large proportion of the known physical elements of the universe. By studying aright the result which we have it in our power to obtain in a watch-glass, we may advance our knowledge of gigantic phenomena, which are now progressing at the bottom of the ocean, or of the wondrous agencies which are in operation, producing light-refracting gems within the secret recesses of the rocky crust of our globe.

CHAPTER VI.

HEAT — SOLAR AND TERRESTRIAL.

Solar and Terrestrial Heat — Position of the Earth in the Solar System — Heat and Light associated in the Sunbeam — Transparency of Bodies to Heat — Heating Powers of the Coloured Rays of the Spectrum — Undulatory Theory — Conducting Property of the Earth's Crust — Convection — Radiation — Action of the Atmosphere on Heat-Rays — Peculiar Heat-Rays — Absorption and Radiation of Heat by dissimilar Bodies — Changes in the Constitution of Solar Beam — Difference between Transmitted and Reflected Solar Heat — Phenomena of Dew — Action of Solar Heat on the Ocean — Circulation of Heat by the Atmosphere and the Ocean — Heat of the Earth — Mean Temperature — Central Heat — Constant Radiation of Heat-Rays from all Bodies — Thermography — Action of Heat on Molecular Arrangements — Sources of Terrestrial Heat — Latent Heat of Bodies — Animal Heat — Natural Phenomena.

We receive heat from the sun, associated with light; and we have the power of developing this important principle in many ways, from nearly every kind of matter. Our convictions are, that the calorific element, whether derived from a solar or terrestrial source, presents no essential difference in its physical characters; but as there are some remarkable peculiarities in the phenomena, as they arise from either one or the other source, it will assist our comprehension of this great principle, if we consider it under the two heads.

Untutored man finds health and gladness in the warmth and light of the sun, and he rears a rugged altar, and bows his soul in prayer, to the principle of fire, which in his ignorance he regards as the giver and supporter of life. The philosopher finds life and organization dependent upon the powers combined in the sunbeam; and, examining the phenomena of this wonderful band of forces, he is compelled to acknowledge that the flame upon the altar is indeed a dim shadow of the infinite wisdom which abides behind the veil.

The present condition of our earth is directly dependent upon the amount of heat we receive from the sun. If it were possible to move this planet so much nearer that orb, that the quantity of heat would be much increased, the circumstance of life would necessarily be so far changed, that nearly all the present races of animals must perish; and the same result would happen from any alteration which threw us yet farther from our central luminary, when, owing to the extremity of cold and the wretchedness of gloom, all living creatures would equally fail to support their organizations.

All things are adapted to the circumstances of the position of the earth in relation to the sun, to which, as we have shown, we are bound by the principle of gravitation; and in our examination it will be found that one common system of harmony runs through all the cosmical phenomena, by which everything is produced that is so beautiful and joyous in this world.

Heat and the other elementary radiant principles, are often combined as the common cause of effects evident to our senses. The warmth of the solar rays and their luminous influence, are not, however, commonly associated in the mind as the results of a single cause. It is only when we come to examine the physical phenomena connected with these radiations that we discover the complexity of the inquiry. Yet it is out of these very subtle researches that we draw the most refined truths. The high inferences to which the analysis of the subtile agencies of creation leads us, render science, pursued in the spirit of truth, a great system of religious instruction.

Although we do not fear that heat and light can be confounded in the mind, so different are their phenomena, yet it is important to show how far these two principles have been separated from each other. Transparent bodies have very various powers of calorific transparency, or transcalescence: some obstructing the heat of bodies at very high temperatures, almost entirely in the thinnest layers; whilst others will allow even the heat of the hand to pass through a thickness of several inches. Liquid chloride of sulphur, which is of a deep red colour, will allow 63 out of 100 rays of heat to pass, and a solution of carmine in ammonia, or a glass stained with oxide of gold, rather a greater number; yet these transparent media obstruct a large quantity of light. Colourless media obstructing scarcely any light, will, on the contrary, prevent the passage of calorific rays. Out of every hundred rays, oil of turpentine will only transmit 31, sulphuric ether 21, sulphuric acid 17, and distilled water only 11. Pure flint glass, however, is permeated by 67 per cent. of the thermic rays, and crown glass by 49 per cent. The most perfect diathermic body is diaphanous salt-rock, which transmits 92, while alum, equally translucent, admits the passage of only 12 per cent. ([43])

Black mica, obsidian, and black glass, are nearly opaque to light, but they allow 90 per cent. of radiant heat to pass through them. Whereas a pale green glass, coloured by oxide of copper, ([44]) covered with a layer of water, or a very thin plate of alum, will, although perfectly transparent to light, almost entirely obstruct the permeation of heat rays.

We thus arrive at the fact that heat and light may be separated from each other; and if we examine the rays of the sun by that analysis which the prism gives us, we shall find that there is no correspondence between intense light and ardent heat. By experiment it has been shown, that where we have in the prismatic spectrum the most light, as in the yellow ray, we have only a temperature of 62° F.; but below the red ray, out of the point of visible light, the temperature is found to be 79°, while at the other end, in the blue ray, it is 56°, and at the end of the violet ray no thermic action can be detected.(45)

From the circumstance, that as we, by artificial means, raise the temperature of any body, and produce intense heat, so we also occasion a manifestation of light after a certain point has been obtained,(46) it has been concluded, somewhat hastily, that heat and light differ from each other only in the rapidity of the undulations of an hypothetical ether.

It must be admitted that the mathematical demonstrations of many of the phenomena of calorific and luminous power, are sufficiently striking to convince us that a wave-movement is common to both heat and light. The undulatory theory, however, requires the admission of so many premises of which we have no proof; its postulates are, indeed, in many cases so gratuitous, that notwithstanding the array of talent which stands forward in its support, we must not allow ourselves to be deceived by the deductions of its advocates, or dazzled by the brilliancy of their displays of learning.

Radiant heat appears to move in waves; but that calorific action is established by any system of undulation, is a deduction without a proof; and the thermic phenomena of matter are more easily explained by the hypothesis of a diffusive subtile fluid.

We have not, however, to prove the correctness of either of the opposing views; indeed, it is acknowledged that many phenomena require for their explanation conditions which are not indicated by either theory.

The earth receives its heat from the sun; a portion of it is *conducted* from particle to particle into the interior of the rocky crust. Another portion produces warmth in the atmosphere around us by *convection*, or the circulation of particles: those warmed by contact with the surface becoming lighter, and ascending to give place to the colder and heavier ones. A third portion is radiated off into space, according to laws which have not been sufficiently investigated, but which are dependent upon the colour, chemical composition, and mechanical structure of the surface.

Few things within the range of our inquiry are more striking than the phenomena of calorific radiation and absorption. They display so perfectly the most refined system of order, and exhibit so strikingly the admirable adaptation of every formation to its particular conditions, and for its part in the great economy of being, that they claim most strongly the study of all who would seek to discover a Poetry in the inferences of science.

Owing to the nature of our atmosphere, we are protected from the influence of the full flood of solar heat. The absorption of caloric by the air has been calculated at about one-fifth of the whole in passing through a column of 6,000 feet. This estimate is, of course, made near the earth's surface; but we are enabled, knowing the increasing rarity of the upper regions of our gaseous envelope in which the absorption is constantly diminishing, to prove, that about one-third of the solar heat is lost by vertical transmission through the whole extent of our atmosphere. ([47])

Experience has proved that the conditions of the sun's rays are not always the same; and there are few persons who have not observed that a more than usual scorching influence prevails under some atmospheric circumstances. This is also evidenced in the effects produced on the foliage of trees, which, though often attributed to electricity, is evidently due to heat. An examination of the solar radiations, as exhibited in the prismatic spectrum, has proved the existence of a class of heat rays, which manifest themselves by a very peculiar deoxidizing power quite independent of their calorific properties. ([48]) We are protected from the severe effects of these rays by the ordinary state of the medium through which the solar heat passes. Our atmosphere is a mixture of gases and aqueous vapour; and it has been found, as already stated, that even a thin film of water, however transparent, prevents the passage of many calorific radiations, and the rays retarded are, for the most part, of that class which have this peculiar scorching power. The air is, in this way, the great equalizer of the solar heat, rendering the earth agreeable to all animals, who, but for this peculiar absorbent medium, would endure, in our temperate clime, the burning rays of a more than African sun.

The surface of the earth during the sunshine — and, in a less degree, even when the sun is obscured by clouds — is constantly receiving heat; but the rate of its absorption varies. Benjamin Franklin showed, by a set of simple but most conclusive experiments, that a piece of black cloth was heated much sooner than cloth of a lighter colour;([49]) and we know, from observations of a similar class, that the bare brown soil receives heat more readily than the bright green grassy carpet of the earth. Consequently, during the winter season, relatively to the quantity poured from its source, heat more easily penetrates the uncovered soil, than during the spring or summer.

There is a constant tendency to an equilibrium; and, during the night, the surface is robbed of more heat by the colder air than by day; and even in these processes of convection and radiation, a similar law prevails to that which is discovered in examining into the rate of calorific absorption.

Every tree spreading its green leaves to the sunshine, or exposing its brown branches to the air — every flower which lends its beauty to the earth — possesses different absorbing and radiating powers. The chalice-like cup of the pure white lily floating on the lake — the variegated tulip — the brilliant anemony — the delicate rose — and the intensely coloured peony or dahlia — have each powers peculiar to themselves for drinking in the warming life-stream of the sun, and for radiating it back again to the thirsting atmosphere. These are no conceits of a scientific dreamer; they are the truths of direct induction; and, by experiments of a simple character, they may be put to a searching test.(50)

A thermometric examination of the various coloured leaves of flowers will readily establish the correctness of the one; and by a discovery of recent date, connected with calorific radiation, which must be particularly described presently, we can, with equal ease and certainty, test the truth of the other.(51)

It follows, as a natural consequence of the position of the sun, as it regards any particular spot on the earth at a given time, that the amount of heat is constantly varying during the year. This variation regulates the seasons. But an analysis of the spectrum shows us that there are some changes regularly taking place in the state of the solar beam, which cannot be referred to the mere alteration of position. It may be inferred, from facts afforded by long-continued observations, that the three classes of phenomena which we detect in the sun's rays are constantly changing their relative proportions. In spring, the chemical agency prevails; in summer, the luminous principle is the most powerful; and in the autumn, the calorific forces are in a state of the greatest activity.(52) The importance of these variations, to the great economy of vegetable life, will be shown when we come to examine the phenomena connected with organization.

A remarkable change takes place in the character of heat in being radiated from material substances. In nature, we often see this fact curiously illustrated. Snow which lies near the trunks of trees or wooden poles, melts much quicker than that which is at a distance from them, — the liquefaction commencing on the side facing the sun, and gradually extending. We see, therefore, that the direct rays of solar heat produce less effect upon the snow than those which are radiated from coloured surfaces. By numerous experiments, it has been shown that these secondary radiations are more abundantly absorbed by snow or white bodies than the direct solar rays themselves. Here is one of the many very curious evidences, which science lays open to us, of the intimate connection between the most ethereal and the grosser forms of matter. Heat, by touching the earth, becomes more earth-like. The subtile principle which, like the spirit of superstition, has the power of passing, unfelt, through the crystal mass, is robbed of its might by embracing the things of earth; and although it still retains the evidences of its refined origin, its movements are shackled as by a clog of clay, and its wings are heavy with the dust of this rolling ball. It has, however, acquired new properties, which fit it for the requirements of creation, and by which its great tasks are facilitated. Matter and heat unite in a common bond, and, harmoniously pursuing the necessities of some universal law, the result is the extension of beautiful forms in every kingdom of nature.

An easy experiment pleasingly illustrates this remarkable change. If a blackened card is placed upon snow or ice in the sunshine, the frozen mass underneath it will be gradually thawed, while that by which it is surrounded, although exposed to the full power of solar heat, is but little disturbed. If, however, we reflect the sun's rays from a metal surface, an exactly contrary result takes place; the uncovered parts are the first to melt, and the blackened card stands high above the surrounding portion.

The evidences of science all indicate the sun as the source, not only of that heat which we receive directly through our atmosphere, but even of that which has been stored by our planet, and which we can, by several methods, develope. We have not to inquire if the earth was ever an intensely heated sphere; — this concerns not our question; as we should, even were this admitted, still have to speculate on the origin — the primitive source of this caloric.

Before, however, we proceed to the examination of the phenomena of terrestrial heat, a few of the great results of the laws of radiation and convection claim our attention.

Nearly all the heat which the sun pours upon the ocean is employed in converting its water into vapour at the very surface, or is radiated back from it, to perform the important office of producing those disturbing influences in the atmosphere, which are essential to the preservation of the healthful condition of the great aërial envelope in which we live.

Currents of air are generally due to the unequal degree in which the atmosphere is warmed. Heat, by expanding, increases the elasticity, and lessens the density, of a given mass. Consequently, the air heated by the high temperature of the tropics, ascends charged with aqueous vapours, whilst the colder air of the temperate and the frigid zones flows towards the equator to supply its place. These great currents of the atmosphere are, independent of the minor disturbances produced by local causes, in constant flow, and by them a uniformity of temperature is produced, which could not in any other way be accomplished. By these currents, too, the equalization of the constituents of the "breath of life" is effected, and the purer oxygen of the "land of the sunny South" is diffused in healthful gales over the colder climes of the North. The waters, too, evaporated from the great central Atlantic ocean, or the far Pacific, are thus carried over the wide-spread continents, and poured in fertilizing showers upon distant lands.

How magnificent are the operations of nature! The air is not much warmed by the radiations of caloric passing from the sun to the earth; but the surface soil is heated by its power of absorbing these rays. The temperature of the air next the earth is raised, and we thus have the circulation of those beneficial currents which are so remarkably regular in the trade winds. A similar circulation, quite independent of the ordinary tidal movement, takes place also in the earth-girdling ocean. The water, warmed by convection from the hot surface of the tropical lands, sets across the Atlantic, from the coast of Africa to the Gulf of Mexico; it is then carried to the shores of Newfoundland, and northwards to the pole. Here we have two immense influences produced by one agency, rendering those parts of the earth habitable and fertile, which but for these great results would sorrow in the cheerless aspect of an eternal winter.

The beautiful phenomena of the formation of dew is also distinctly connected with the peculiar properties which we have been studying. When from the bright blue vault of heaven, the sparkling constellations shower their mild light over the earth, the flowers of the garden and the leaves of the forest become moist with a fluid of the most translucid nature. Well might the ancients imagine that the dews were actually shed from the stars; and the alchemists and physicians of the Middle Ages conceive that this pure distillation of the night possessed subtile and penetrating powers beyond most other things; and the ladies of those olden times endeavour to preserve their charms in the perfection of their youthful beauty through the influences of washes procured from so pure a source.([53])

Science has removed the veil of mystery with which superstition had invested the formation of dew; and in showing to us that it is a condensation of vapour upon bodies according to a fixed law of radiation, it has also developed so many remarkable facts connected with the characters of material creations, that a much higher order of poetry is opened to the mind than that which, though beautiful, sprang merely from the imagination.

Upon the radiation of heat depends the formation of dew, and bodies must become colder than the atmosphere before it will be deposited upon them. Different substances, independent of colour, have the property of projecting heat from their surfaces with different degrees of force. Rough and porous surfaces radiate heat more rapidly than smooth ones, and are consequently reduced in temperature; and, if exposed, covered with dew sooner than smooth and dense bodies are. The grass parterre glistens with dew, whilst the hard and stony walk is unmoistened.([54])

Colourless glass is very readily suffused with dampness, but polished metals are not so, even when dews are heavily condensed on other bodies. To comprehend fully the phenomena of the formation of dew, we must remember that the entire surface of the earth is constantly radiating heat into space; and that, as by night no absorption of caloric is taking place, it naturally cools.([55]) As the substances spread over the earth become colder than the air, they acquire the power of condensing the vapour with which the atmosphere is always charged. The bodies which cover this globe are very differently constituted; they possess dissimilar radiating powers, and consequently present, when examined by delicate thermometers, varying degrees of temperature. By the researches of Dr. Wells,([56]) which may be adduced as an example of the best class of inductive experiments, we learn that the following differences in sensible heat were observed at seven o'clock in the evening:—

The air four feet above the grass	60¾
Wool on a raised board	54½
Swandown on ditto	53
The surface of the raised board	57
Grass plat	51

Dew is most abundantly deposited on clear, calm nights, during which the radiation from the surface of the earth is uninterrupted. The increased cold of such nights over those obscured by clouds is well known. The clouds, it has been proved, act in the same way as the screens used by gardeners to protect their young plants from the frosts of the early spring, which obstruct the radiation, and, in all probability, reflect a small quantity of heat back to the earth.

It is not improbable that the observed increase in grass crops, when they have been strewn with branches of trees or any slight shades, may be due to a similar cause.([57])

There are many remarkable results dependent entirely on the colours of bodies, which are not explicable upon the idea of difference in mechanical arrangement. We know that different colours are regulated by the powers which structures have of absorbing and reflecting light; consequently ,a blue surface must have a different order of molecular arrangement from a red one. But there are some physical peculiarities which also influence calorific radiation, quite independently of this *surface* condition. If we take pieces of red, black, green, and yellow glass, and expose them when the dew is condensing, we shall find that moisture will show itself first on the yellow, then on the green glass, but that none will appear on either the black or red glasses. The same thing takes place if we expose coloured fluids in white glass bottles or troughs, in which case the surfaces are all alike. If against a sheet of glass, upon which moisture has been slightly frozen, we place similarly coloured glasses to those already described, it will be found that the earliest heat-rays will so warm the red and the black glasses, that the ice will be melted opposite to them, long before any change will be seen upon the frozen film covered by the other colours.

The order in which heat permeates coloured media, it has already been shown, very nearly agrees with their powers of radiation.

These most curious results have engaged the attention of Melloni, to whose investigations we owe so much; and from the peculiar order of radiations, which present phenomena of an analogous character to those of the coloured rays of light, obtained by him from dissimilarly coloured bodies, he has been led to imagine the existence of a "heat colouration." That is — the heat-rays are supposed to possess properties like luminous colour, although invisible; and, consequently, that a blue surface has a strong affinity for the blue heat-rays, a red surface for the red ones, and so on through the scale. The ingenuity of this hypothesis has procured it much attention, and it is valuable as one of the aids to the discovery of those truths which science so earnestly seeks to reveal. ([58])

Can anything be more calculated to impress the mind with the consciousness of the high perfection of natural phenomena, than the fact, that the colour of a body should powerfully influence the transmission of a principle which is diffused through all nature, and also determine the rate with which it is to pass off from its surface. Some recent experiments have brought us acquainted with other facts connected with these heat-radiations, and the power of caloric, as influenced by the colourific rays, to produce molecular changes in bodies, which bear most importantly on our subject.

If we throw upon a plate of polished metal a prismatic spectrum (deprived, as nearly as possible, of its chemical power, by being passed through a deep yellow solution — which possesses this property in a very remarkable manner, as will be explained when we come to the examination of the chemical action of the sun's rays) — it will be found, if we afterwards expose the plate to the action of vapour, very slowly raised from mercury, that the space occupied by the red rays, and those which lie without the spectrum below it, will condense the vapour thickly, while the portion corresponding with the other rays will be left untouched. This affords us evidence of the power of solar heat to produce very readily a change in the molecular structure of solid bodies. If we allow the sun's rays to permeate coloured glasses, and then fall upon a polished metallic surface, the result, on exposing the plate to vaporization, will be similar to that just described. Under yellow and green glasses no vapour will be condensed; but on the space on which the rays permeating a red glass, or even a blackened one, fall, a very copious deposit of vapour will mark with distinctness the spaces these glasses covered. More remarkable still, if these or any other coloured bodies are placed in a box, and a polished metal plate is suspended a few lines above them, the whole being kept *in perfect darkness* for a few hours, precisely the same effect takes place as when the arrangement is exposed to the full rays of the sun. Here we have evidence of the radiating heat of bodies, producing even in darkness the same phenomena as the transmitted heat-rays of the sun. We must, however, return to the examination of some of these and other analogous influences under the head of actino-chemistry.

From these curious discoveries of inductive research, we learn some high truths. Associated with light — obeying many of the same laws — moving in a similar manner — we receive a power which is essential to the constitution of our planet. This power is often manifested in such intimate combination with the luminous principle of the solar rays, that it has been suspected to be but another form of the same agency. While, however, we are enabled to show the phenomena of the one without producing those which distinguish the other, we are constrained to regard heat as something dissimilar to light. It is true that we appear to be tending toward some point of proof on this problem; but we are not in a position to declare them to be forms of one common power, or "particular solutions of one great physical equation."([53]) In many instances it would certainly appear that one of these forces was directly necessary to the production of the other; but we have also numerous examples in which they do not stand in any such correlation.

We learn, from the scientific facts which we have been discussing, a few of the secrets of natural magic. In their relations to heat, every flower, which adds to the adornment of the wilds of nature or the carefully-tended garden of the florist, possesses a power peculiar to itself; and, as we have before indicated, the

"Naiad-like lily of the vale,"

and,

"——The pied wind-flowers, and the tulip tall,
And narcissi, the fairest among them all,"

are, by their different colours, prevented from ever having the same temperatures under the same sunshine.

Every plant bears within itself the measure of the caloric which is necessary for its well-being, and is endued with functions which mutely determine the relative amount of dew which shall wet its coloured leaves. Some of the terrestrial phenomena of this remarkable principle will still further illustrate the title of this volume.

To commence with the most familiar illustrations, let us consider the consequences of change of temperature. However slight the additional heat may be to which a body is subjected, it expands under its influence; consequently, every atom which goes to form the mass moves under the excitation. The differences between the temperature of day and night are considerable; therefore all bodies expand under the influence of the higher, and contract under that of the lower, temperature. During the day, any cloud obscuring the sun produces, in every solid, fluid, or aëriform body, within the range of solar influence, a check: the particles which had been expanding under the force of heat suddenly contract. Thus there must of necessity be, during the hours of sunshine, a tendency in all bodies to dilate, and during the hours of night they must be resuming their original conditions.

Not only do dissimilar bodies radiate heat in different degrees, but they conduct it also with constantly varying rates. Caloric passes along silver or copper with readiness, compared to its progress through platinum. It is conducted by glass but slowly, and still more slowly by wood and charcoal. The metallic oxides or earths are bad conductors of heat, by which provision the caloric absorbed by the sun's rays is not carried away from the surface of this planet so rapidly as it would have been had it been of metal; but is retained in the superficial crust to produce the due temperature for healthful germination and vegetable growth. The wool and hair of animals are still inferior conductors, and thus, under changes of climate and of seasons, the beasts of the field are secured against those violent transitions from heat to cold which would be fatal to them. Hair is a better conductor than wool: hence, by nature's alchemy, hair is changed into wool on the approach of winter, and feathers into down.

It is, therefore, evident that the rate at which solar heat is conducted into the crust of the earth must alter with the condition of the surface upon which it falls. The conducting power of all the rocks which have been examined, is found to vary in some degree.([8])

It follows, as a natural consequence of the position of the sun to the earth, that the parts near the equator become more heated than those remote from it. As this heat is conducted into the interior of the mass, it has a tendency to move to the colder portions of it, and thus the heat absorbed at the equator flows towards the poles, and from these parts is carried off by the atmosphere, or radiated into space. Owing to this, there is a certain depth beneath the surface of our globe at which an equal temperature prevails, the depth increasing as we travel north or south from the equator.(61)

A question of great interest, in a scientific point of view, is the calorific condition of the centre of the earth. We are, of course, without the means of solving this problem; but we advance a little way onwards in the inquiry by a careful examination of subterranean temperature at such depths as the enterprise of man enables us to reach. These researches show us, that where the mean temperature of the climate is 50°, the temperature of the rock at 59 fathoms from the surface is 60°; at 132 fathoms it is 70°; at 239 fathoms it is 80°: being an increase of 10° at 59 fathoms deep, or 1° in 35.4 feet; of 10° more at 73 fathoms deeper, or 1° in 43.8 feet; and of 10° more at 114 fathoms still deeper, or 1° in 64.2 feet.(62)

Although this would indicate an increase to a certain depth of about one degree in every fifty feet, yet it would appear that the rate of increase diminishes with the depth, and that the heat of the earth, so far as man can examine it, is due to the absorption of the solar rays by the surface. The mean annual temperature of this planet is of course subject to great variations: at the equator we may regard it as uniformly existing at 80°, while at the poles it is below the freezing point of water; and as far as observations have been made, the subterranean temperatures bear a close relation to the thermic condition of the climate of the surface. The circulation of water through faults or fissures in the strata is, without doubt, one means of carrying heat downwards much quicker than it would be conducted by the rocks themselves. It is now, however, found that the quantity of water increases with the depth. In the mines of Cornwall, unless where the ground is very loose, miners find that, after about 150 fathoms

(900 feet), the quantity of water rapidly diminishes. That water must ascend from very much greater depths is certain, from the high temperatures at which many springs flow out at the surface. At the bottom of the United Mines in Cornwall, water rises from one part of the lobe at 90° ; and one of the levels in these workings is so hot that, notwithstanding a stream of cold water is purposely brought into it to reduce the temperature, the miners work nearly naked, and will bathe in water at 80° to cool themselves. At the bottom of Tresavean Mine, in the same county, about 320 fathoms from the surface, the temperature is nearly 100°.

One cause of the great heat of many of our deep mines, which appears to have been entirely lost sight of, is the chemical action going on upon large masses of pyritic matter in their vicinity. The heat, which is so oppressive in the United Mines, is, without doubt, due to the decomposition of immense quantities of the sulphuret of iron, known to be in this condition at a short distance from these mineral works.

As a proof that the heat, which we are enabled to measure beneath the earth's surface, is due to the conducting powers of the rocks themselves, it has been found that the line of equal temperature follows, as nearly as possible, the elevations and depressions which prevail upon the surface.

Whether or not the subterranean bands of equal heat have any strict relation, upon a large scale, to the isothermic lines which have been traced around most portions of our globe, is a point which has not yet been so satisfactorily determined as to admit of any general deductions.

The Oriental story-teller makes the inner world a place of rare beauty — a cavern temple, bestudded with self-luminous gems, in which reside the spiritual beings to whom the direction of the inorganic world is confided.

Man, in the height of his knowledge, has had dreams as absurd as this; and amid the romances of science, there are not to be found any more strange visions than those which relate to the centre of our globe. At the same time it must be admitted, that many of the peculiar phenomena which modern geological researches have brought to light, are best explained on the hypothesis of a cooling mass, which necessarily involves the existence of a very high temperature towards the centre.

We have already noticed some remarkable differences between solar and terrestrial heat; but a class of observations by Delaroche([63]) still requires our attention. Solar heat passes freely through colourless glass, whereas the radiations from a bright fire or a mass of incandescent metal are entirely obstructed by this medium. If we place a lamp or a ball of glowing hot metal before a metallic reflector, the focus of accumulated heat is soon discovered; but if a glass mirror is used, the light is reflected, but not the heat; whereas, with the solar rays, but little difference is detected, whether vitreous or metallic reflectors are employed. It is well known that glass lenses refract both the light and heat of the sun, and they are commonly known as burning-glasses: the heat accumulated at their focal point being of the highest intensity. If, instead of the solar beam, we employ, in our experiments, an intense heat produced by artificial means, the passage of it is obstructed, and the most delicate thermometers remain undisturbed in the focus of the lens. Glass exposed in front of a fire becomes warm, and by conduction the heat passes through it, and a secondary radiation takes place from the opposite side.([64]) It has been found that glass is transcalescent, or diathermic, to some rays of terrestrial heat, and adiathermic, or opake for heat, to others([65]) — that the capability of permeating glass increases with the temperature of the ignited body — and that rays which have passed one screen traverse a second more readily. It would, however, appear that something more than a mere elevation of temperature is necessary to give terrestrial calorific radiations the power of passing through glass screens, or, in other words, to acquire the properties of solar heat.

To give an example: The heat of the oxy-hydrogen flame is most intense, yet glass obstructs it, although it may be assisted by a parobolic reflector. If this flame is made to play upon a ball of lime, by which a most intense light is produced, the heat, which has not been actually increased, acquires the power of being refracted by a glass lens, and combustible bodies may be ignited in its focus.

It certainly appears from these results that the undulatory hypothesis holds true, so far as the motion of the calorific force is concerned. At a certain rate, the vibrations are thrown back or stopped by the opposing body, while in a state of higher excitation, moving with increased rapidity, they permeate the screen. ([66]) This does not, indeed, interfere with the refined theory of Prévost,([67]) which supposes a mutual and equal interchange of caloric between all bodies.

The most general effect of heat is the expansion of matter; solids, liquids, and airs, all expand under its influence. If a bar of metal is exposed to calorific action, it increases in size, owing to its particles being separated farther from each other: by continuing this influence, after a certain time the cohesion of the mass is so reduced that it melts, or becomes liquid, and, under the force of a still higher temperature, this molten metal may be dissipated in vapour. It would appear as if, under the agency of the heat applied to a body, its atoms expanded, until at last, owing to the tenuity of the outer layer or envelope of each atom, they were enabled to move freely over each other, or to interpenetrate without difficulty. That heat does really occasion a considerable disturbance in the corpuscular arrangement of bodies, may be proved by a very interesting experiment. A bar of heated *metal is placed to cool, with* one end supported upon a wedge of another sort of metal, the other resting on the ground. In cooling, a distinct musical sound is given out, owing to the vibratory action set up among the particles of matter moving as the temperature declines. (68)

Heat is diffused through all bodies in nature, and, as we shall presently see, may be developed in many different ways. We may, therefore, infer, that in converting a sphere of ice into water, and that again into steam, we have done nothing more than interpenetrate the mass with a larger quantity of caloric, by which its atoms are more widely separated, and that thus its molecules become more ductile and elastic. Thus, from a solid state, the water becomes fluid ; and then, if the expansive force is continued, an invisible vapour. If these limits are passed by the powers of any greatly increased thermic action, the natural consequence, it must be seen, will be the separation of the atoms from each other, to such an extent that the molecule is destroyed, and chemical decomposition takes place.

By the agency of the electricity of the voltaic battery, we are enabled to produce the most intense heat with which we are acquainted, and by a peculiarly ingenious arrangement, Mr. Grove has succeeded in resolving water into its constituent elements — oxygen and hydrogen gases. That this decomposition is not due to the voltaic current was subsequently proved by employing platina, heated by the oxy-hydrogen flame. (69)

This interesting question has been examined with great care by Dr. Robinson of Armagh, who has shown that, as the temperature of water is increased, the affinity of its elements is lessened, until at a certain point it is eventually destroyed. This new and startling fact appears scarcely consistent with our knowledge, that a body heated so as to be luminous has the power of causing the combination of the elements of water with explosive violence.

(70) But as this acute experimental philosopher somewhat boldly but still most reasonably suggests: "Is it not probable that, if not light, some other actinic power (like that which accompanies light in the spectrum, and is revealed to us by its chemical effects in the processes of photography) is evolved by the heat, and, though invisible, determines, in conjunction with the affinity, that atomic change which transforms the three volumes of oxygen and hydrogen into two of steam?"(71)

This speculation explains, in a very satisfactory manner, some results which were obtained by Count Rumford, in 1798. In a series of experiments, instituted for the purpose of examining "those chemical properties of light which have been attributed to it," he has shown that many cases of chemical decomposition occur in perfect darkness, under the influence of heat, which are precisely similar to those produced by exposure to the sun's rays.(72)

It must, however, be remembered, that both solar light and heat are sometimes found in direct antagonism to actinic power, and that the most decided chemical changes are produced by those rays in which neither heat nor light can be detected. The remarkable phenomena of this class will be explained under the head of actinism.

One of the most curious relations which, as yet, have been discovered between light and heat is, that the temperature of the incandescence of all bodies, excepting such as are phosphorescent, is uniform. This point on the thermometer (Fahrenheit's scale) may probably be regarded as, or very near, 1000°, when the eye by perfect repose is enabled to detect the first luminous influence. Daniel has fixed this point at 980°, Wedgwood at 947°, and Draper at 977°.(73) Dr. Robinson and Dr. Draper, by independent observations, have both arrived at the conclusion, that the first gleam of light which appears from heated platina is not red, but of a lavender gray, the same in character as that detected by Sir John Herschel among the most refrangible rays of the solar spectrum.(74)

It must be admitted, that the question of the identity, or otherwise, of light and radiant heat, is beset with difficulties. Many of their phenomena are very similar — many of their modes of action are alike: they are often found as allied agencies; but they as frequently exhibit extreme diversity of action, and they may be separated from each other.

We have now examined the physical conditions and properties of this most important element, and we must proceed to learn something of the means by which it may be developed, independently of its solar source.

This extraordinary principle exists in a latent state in all bodies, and may be pressed out of them. The blacksmith hammers a nail until it becomes red-hot, and from it he lights his match; the iron has by this process become more dense, and percussion will not again produce incandescence until the bar has been exposed in fire to a red heat. The Indian produces a spark by the attrition of two pieces of wood. By friction, two pieces of ice may be made to melt each other; and could we, by mechanical pressure, force water into a solid state, an immense quantity of caloric would be set free. By the condensation of hydrogen and oxygen gases, pulverulent platinum will become glowing red-hot, and, with certain precautions, even the compact metal, platinum, itself: the heat being derived from the gases, the union of which it has effected. A body passing from the solid to the fluid state absorbs heat from all surrounding substances, and hence a degree of cold is produced. The heat which is thus removed is not destroyed — it is held combined with the fluid; it exists in a latent state. Fluids, in passing into a gaseous form, also rob all surrounding bodies of an amount of heat necessary to maintain the aëriform condition. From the air or from the fluid this heat may, as we have shown above, be again extracted. Locked in a pint measure of air, there exists sufficient caloric to raise several square inches of metal to a glowing red heat. By the compression of atmospheric air this may be shown, and with a small condensing syringe a sufficient quantity of heat may be set free to fire the *Boletus igniarius*, which, impregnated with nitre, is known as *amadou*. We are acquainted with various sources of heat for artificial purposes: the flint and steel, and the modern Lucifer-match, are the most common. These of themselves would admit of a lengthened discourse; but it is necessary that we carefully examine some of the less familiar phenomena of heat under the influences of changes of chemical condition.

If spirits of wine and water are mixed together, a considerable degree of heat is given out, and by mixing sulphuric acid and water, an infinitely larger amount. If oil of vitriol and spirits of wine, or aquafortis (nitric acid) and spirits of turpentine, at common temperatures, be suddenly mixed, so much heat is set free as to ignite the spirits. In all these instances there is a condensation of the fluid. In nearly all cases of solution, cold is produced by the absorption of the heat necessary to sustain the salt in a liquid form; but when potash dissolves in water, heat is given out, which is a fact we cannot yet explain. If potassium is placed on water, it sets fire, by the heat produced, to the hydrogen gas liberated from it. Antimony and many other metals thrown into chlorine gas ignite and burn with brilliancy: the same phenomenon takes place in the vapours of iodine or bromine. Many chemical combinations, as the chlorate of potash and sulphur, explode with a blow; whilst the slightest friction occasions the detonation of the fulminating salts of silver, mercury, and gold. Compounds of nitrogen and chlorine, or iodine, are still more delicately combined — the former exploding with fearful violence on the contact of an oleaginous body, and the latter with the smallest elevation of temperature: both of them destroying the vessels in which they may be contained. These fearful disturbances of combination can only be explained upon the supposition, that the particles have the property of condensing around them an enormous quantity of the calorific and chemical principle, and retaining them in a latent state until some disturbance renders them sensible, by which the sudden destruction of the chemical union is produced, and the full powers of heat and actinism are developed. The fact of great heat being evolved during the conversion of a body from a solid to a gaseous state, which is a striking exception to the law of latent heat, as it prevails in most cases, admits of no more satisfactory explanation.

As mechanical force produces calorific excitation, so we find that every movement of sap in vegetables, and of the blood and fluids in the animal economy, causes a sensible increase of heat. The chemical processes constantly going on in plants and animals are another source of heat; and to nervous energy and to muscular movement, must we also look for the sustaining caloric which is essential to the health and life of the latter. Digestion has been considered as a process of combustion; and the action between the elements of food, and the oxygen conveyed by the circulation of the blood to every part of the body, regarded as the source of animal heat; and, without doubt, it is one great source, although it cannot be regarded as the only one.([75])

The *vis vitæ*, or vital power, influences the delicate and beautiful system of nerves; and as life runs through them, from the brain to the extremities of the members of the body, an essence of the rarest and most subtile order, a diffusive influence, it sets those tender threads in rapid vibration, and heat is developed. By this action, the circulation of the blood is effected; the muscle is maintained in an elastic condition, ready to perform the tasks of the will; and through these agencies is the warm and fluid blood fitted to receive its chemical restoratives in the lungs, and the stomach to support changes to which it is designed — chemical also — by which more heat is liberated. Was digestion — eremacausis, as the slow combustion produced by combination with oxygen is called — the only source of animal heat, why should the injury of one filmy nerve place a member of the body for ever in the condition of stony coldness? Or why, chemical action being most actively continued after a violent death, by the action of the gastric juices upon the animal tissues, should not animal heat be maintained for a much longer period than it is found to be?([76])

In studying the influences of caloric upon the conditions of matter, we must regard the effects of extreme heat, and also of the greatest degrees of cold which have been obtained.

There are a set of experiments by the Baron Cagniard de la Tour, which appear to have a very important bearing on some conditions that may be supposed to prevail in nature, particularly if we adopt the view of a constantly increasing temperature towards the centre of our earth. If water, alcohol, or ether, is put into a strong glass tube of small bore, and, the ends being hermetically sealed, the whole is exposed to a strong heat, the fluid disappears, being converted into a transparent gas; but, upon cooling, it is again condensed, without loss, into its original fluid state.([77]) In this experiment, fluid bodies have been converted into elastic transparent gases without any change of volume, under the pressure of their own atmospheres. We can readily conceive a similar result occurring upon a far more extensive scale. In volcanic districts, at great depths, and consequently under the pressure of the superincumbent mass, the siliceous rocks, or even metals, may, from the action of intense heat, be brought into a gaseous or fluid condition without any change of volume, since the elastic force of heat is opposed by the rigid resistance of the pressure of the surrounding rocks.

Directly connected with these results of Cagniard de la Tour, are a yet more remarkable set of phenomena, which have been investigated by M. Boutigny,(78) and generally known as the "spheroidal condition" of bodies. If water is projected upon hot metal, it instantly assumes a spheroidal form — an internal motion of its particles may be observed — it revolves with rapidity, and evaporates very slowly. Even if a silver or platinum capsule, when brought to a bright red heat, is filled with cold water, the whole mass assumes the spheroidal state, the temperature of the fluid constantly remaining considerably below the boiling point, so long as the red heat is maintained. If we allow the vessel to cool below redness, in the dark, the water bursts into active ebullition, and is dissipated into vapour with almost explosive violence.

Another form of this experiment is exceedingly instructive. If a mass of white hot metal is suddenly plunged into a vessel of cold water, the incandescence is not quenched, the metal shines with a bright white light, and the water is seen to circulate around, but at some distance from the glowing mass, being actually repelled by the calorific agency. At length, when the metal cools, the water comes in contact with it, and boils with energy.

A result similar to this was observed by Perkins, but its correctness most unjustly doubted. Having made an iron shell, containing water, red-hot, he caused a hole to be drilled into it, and he was surprised to find that no water flowed through the orifice until the iron was considerably cooled, when it issued forth with violence in the form of steam. If water is poured upon an iron sieve, the wires of which are made red-hot, it will not percolate; but on cooling it will run through rapidly. M.

Boutigny, pursuing this curious inquiry, has recently proved that the moisture upon the skin is sufficient to protect it from disorganization, if the arm is rapidly plunged into baths of melted metal. The resistance of the surfaces is so great, that little elevation of temperature is experienced.([79])

We have now seen that heat appears to produce chemical composition — that it decomposes combined elements — that it alters the conditions of bodies, and actually maintains so powerfully a repellent force, that fluids cannot touch the heated body. More than this, it exerts a most powerful influence over all chemical relations. If, to give one example, the volatile element iodine is put into a glowing-hot capsule, it resolves itself immediately into a spheroid. Potash rapidly combines with iodine; but if a piece of this alkali is thrown upon it in the capsule, it also takes the spheroidal form, and both bodies revolve independently of each other, their chemical affinities being entirely suspended; but allow the capsule to cool, and they combine immediately.

These experiments of Cagniard de la Tour and of Boutigny (d'Evreux), connect themselves, in a striking manner, with those of Mr. Grove and Dr. Robinson; and they teach us that but a very slight alteration in the proportions of the calorific principle given to this planet would completely change the character of every material substance of which it is composed, unless there was an alteration in the physical condition of the elements themselves.

Supposing the ordeal of fiery purification to take place upon this planet, these experiments appear to indicate the mighty changes which would thence result. There would be no annihilation, but everything would be transformed from the centre of the globe to the verge of its atmosphere — old things would pass away, all things become new, and the beautiful mythos of the phœnix be realized in the fresh creation.

The deductions to be drawn from the results obtained by abstracting heat from bodies are equally instructive. By taking advantage of the cooling produced by the rapid solution of salts of several kinds in water, an intense degree of coldness may be produced.([80]) Indeed, the absorption of heat by liquefaction may be shown by the use of metallic bodies alone. If lead, tin, and bismuth are melted together, and reduced to a coarse powder by being poured into water, and the alloy then dissolved in a large quantity of quicksilver, the thermometer will sink nearly 50 degrees. An intense amount of cold will result from the mixture of muriate of lime and snow, by which a temperature of 50° below the zero of Fahrenheit, or 82° below the freezing point of water, is produced. By such a freezing mixture as this, mercury will be rendered solid. A degree of cold, however, far exceeding it, has lately been obtained by the use of solid carbonic acid and ether.([81]) Solid carbonic acid is itself procured from the gas liquefied by pressure; which liquid, when allowed to escape into the air, evaporates so rapidly that a large quantity of it is congealed by being robbed of its combined heat by the vaporizing portion. When this solid acid is united with ether, a bath is formed in which the carbonic acid will remain solid for twenty or thirty minutes. By a mixture of this kind, placed under the receiver of an air-pump, a good exhaustion being sustained, a degree of cold, 166° below zero, is secured. By this intense cold, many of the bodies which have hitherto been known to us only in the gaseous state, have been condensed into liquids and solids. Olefiant gas, a compound of hydrogen and carbon, was brought into a liquid form. Hydriodic and hydrobromic acids could be condensed into either a liquid or a solid form. Phosphoretted hydrogen, a gas which inflames spontaneously when brought into contact with the air or with oxygen, became a transparent liquid at this great reduction of temperature. Sulphurous acid may be condensed, by pressure and a reduction of temperature, into a liquid which boils at 14° Fahrenheit, but by the carbonic acid bath it is converted into a solid body, transparent and without colour. Sulphuretted hydrogen gas solidifies at 122° below zero, and forms a white substance resembling a mass of crystals of sea-salt.

A combination of the two gases, chlorine and oxygen, becomes solid at — 75°, and the protoxide of nitrogen at — 150°. Cyanogen, a compound of carbon and nitrogen, the base of Prussic acid, is solidified at 30° below the zero of our thermometric scale. The well-known pungent compound, ammonia, so exceedingly volatile at common temperatures, is converted into a crystalline, translucent, white substance at the temperature of — 103°. The difficulties which necessarily attend the exposure of a body to extreme cold and great pressure at the same time, appear to be the only obstacle to the condensation of oxygen, hydrogen, and nitrogen gases. A sufficient amount of condensation was, however, effected by Dr. Faraday, to lead him to the conclusion, arrived at also by other evidences, that hydrogen, the lightest of the ponderable bodies, partakes of the nature of a metal. ([82])

The refinements of Grecian philosophy saw, without the aids of inductive science, that the outward vesture of nature covered a host of mysterious agencies to which its characteristics were directly due. In their dream of the four elements, fire, the external and visible form of heat, was regarded as the cause of vitality, and the disposer of every organized and unorganized condition of matter. Their idealizations have assumed another form, but the researches of modern science have only established their universality and truth.

The great agents at work in nature — the mighty spirits bound to never-ending tasks, which they pursue with unremitting toil, are of so refined a character, that they will probably remain for ever unknown to us. The arch-evocator, with the wand of induction, calls; but the only answer to his evocation is the manifestation of power in startling effects. Science pursues her inquiries with zeal and care: she tries and tortures nature to compel her to reveal her secrets. Bounds are, however, set to the powers of finite search: we may not yet have reached the limits within which we are free to exercise our mental strength; but, these limits reached, we shall find an infinite region beyond us, into which even conjecture wanders eyeless and aimless, as the blind Cyclops, groping in his melancholy cave. ([83])

All we know of heat is, that striking effects are produced which we measure by sensation, and by instruments upon which we have observed that given results will be produced under certain conditions: of anything approaching to the cause of these, we are totally ignorant. The wonder-working mover of some of the grandest phenomena in nature — giving health to the organic world, and form to the inorganic mass — producing genial gales and dire tornadoes — earthquake strugglings and volcanic eruptions — ministering to our comforts in the homely fire — and to advancement in civilization in the mighty furnace, and the ingenious engine which drains our mines, or traverses our country with bird-like speed, — will, in all probability, remain for ever unknown to man. The immortal Newton, many of whose guesses have a prophetic value, thus expresses himself: "Heat consists in a minute vibratory motion in the particles of bodies, and this motion is communicated through an apparent vacuum by the undulations of a very subtile elastic medium, which is also concerned in the phenomena of light."

Our experimental labours and our mathematical investigations, have considerably advanced our knowledge since the time of Newton; yet still, each theory of heat strangely resembles the mystic lamp which the Rosicrucian regarded as a type of eternal life — a dim and flickering symbol, in the tongue-like flame of which imagination, like a child, can conjure many shapes.

CHAPTER VII.

LIGHT.

Theories of the Nature of Light — Hypotheses of Newton and Huyghens — Sources of Light — The Sun — Velocity of Light — Transparency — Dark lines of the Spectrum — Absorption of Light — Colour — Prismatic Analysis — Rays of the Spectrum — Rainbow — Diffraction — Interference — Goëthe's Theory — Polarization — Magnetization of Light — Vision — The Eye — Analogy — Sound and Light — Influence of Light on Animals and Vegetables — Phosphorescence arising from several causes — Artificial Light — Its Colour dependent on Matter.

Light, the first creation, presents to the inquiring mind a series of phenomena of the most exalted character. The glowing sunshine, painting the earth with all the brilliancy of colour, and giving to the landscape the inimitable charm of every degree of illumination from the gray shadow to the golden glow; — the calm of evening, when, weary of the "excess of splendour," the eye can repose in tranquillity upon the "cloud-land" of the west, and watch the golden and the ruddy hues fade slowly into the blue tincture of night; — and the pale refulgence of the moon, with the quiet sparkle of the sun-lit stars, all tend to impress upon the soul the great truth, that, where there is light, organization and life are found, and beyond its influence death and silence hold supreme dominion. ([84]) Through all time we have evidences that this has been the prevailing feeling of the human race, derived, of course, from their observation of the natural phenomena dependent upon luminous agency. In the myths of every country, impersonations of light prevail, and to these are referred the mysteries of the perpetual renewal of life on the surface of the earth.

This presentiment of a philosophic truth, in the instance of the poet sages of intellectual Greece, was advanced to the highest degree of refinement; and the sublime exclamation of Plato: "Light is truth, and God is light," approaches nearly to a divine revelation.

As the medium of vision — as the cause of colour — as a power influencing in a most striking manner all the forms of organization around us, light presented to the inquiring minds of all ages a subject of the highest interest.

The ancient philosophers, although they lost themselves in the metaphysical subtleties of their schools, could not but discover in light an element of the utmost importance in natural operations. The alchymists regarded the luminous principle as a most subtile fluid, capable of interpenetrating and mingling with gross matter: gold being supposed to differ from the baser metals only in containing a larger quantity of this ethereal essence. (85) Modern science, after investigating most attentively a greater number of the phenomena of light, has endeavoured to assist the inquiry by the aid of hypotheses. Newton, in a fine theory which exhibits the refined character of that great philosopher's mind, supposes luminous particles to dart from the surfaces of bodies in all directions — that these infinitely minute particles are influenced by the attracting and repelling forces of matter, and thus turned back, or reflected, from their superficies in some cases, and absorbed into their interstitial spaces in others.

Huyghens, on the contrary, supposes light to be caused by the waves or vibrations of an elastic medium diffused through all space, which waves are propagated in every direction from the luminous body. In the one case, a luminous particle is supposed actually to come from the sun to the earth; in the other, the sun only occasions a disturbance of the *ether*, which extends with great rapidity, in the same manner as a wave spreads itself over the surface of a lake.

Nearly all the facts known in the time of Newton, and those discovered by him, were explained most satisfactorily by his theory; but it was found they could be interpreted equally well by the undulatory hypothesis, with the exception of the production of colour by prismatic refraction. Although the labours of the most gifted minds have been given, with the utmost devotion, to the support of the vibratory theory, this simple fact has never yet received any satisfactory explanation from it; and there are numerous discoveries connected with the molecular and chemical disturbances produced by the sun's rays, of which its ardent supporters do not even attempt an explanation.

In both theories, a wave motion is admitted, and every fact renders it probable that this mode of progression applies not only to light, but to the so-called *imponderable forces.* Admitting, therefore, the undulatory movement of luminous rays, we shall not stop to consider those points of the discussion which have been so ably dealt with by Young, Laplace, Fresnel, Biot, Fraunhofer, Herschel, Brewster, and others, but proceed at once to consider the sources of light, and its more remarkable phenomena. ([86])

The sun is the greatest permanently luminous body we are acquainted with, and that orb is continually pouring off light from its surface in all directions at the rate, through the resisting medium of space and of our own atmosphere, of 192,000 miles in a second of time. It has been calculated, however, that it would move through a vacuum with the speed of 192,500 miles in the same period. We, therefore, learn that a ray of light requires eight minutes and thirteen seconds to come from the sun to us. In travelling from the distant planet Uranas, nearly three hours are exhausted; and from the nearest of the fixed stars each ray of light requires more than six years to traverse the intervening space between them and the earth. Allow the mind to advance to the regions of the nebulæ, and it will be found that hundreds of years must glide away during the passage of their radiations. Consequently, if one of those masses of matter, or even one of the remote fixed stars, was "blotted out of heaven" to-day, several generations of the finite inhabitants of this world would fade out of time before the obliteration could be known to man. Here the immensity of space assists us in our conception, limited though it be, of the for ever of eternity. ([87])

All the planets of our system shine with reflected light, and the moon, our satellite, also owes her silvery lustre to the sun's radiations. The fixed stars are, in all probability, suns shining from the far distance of space, with their own self-emitted lights. By the photometric researches of Dr. Wollaston we learn, however, that it would take 20,000 millions of such orbs as Sirius, the brightest of the fixed stars, to afford as much light as we derive from the sun. The same observer has proved that the brightest effulgence of the full moon is yet 801,072 times less than the luminous power of our solar centre.

Chemical action is also a source of light; and, under several circumstances in which the laws of affinity are strongly exerted, a very intense luminous effect is produced. In the electric spark we have the development of light; and the arc which is formed between the poles of a powerful voltaic battery, affords us the most intense artificial illumination with which we are acquainted. In addition to these, we have the peculiar phenomena of phosphorescence arising from chemical, calorific, electrical, actinic, and vital excitation, all of which must be particularly examined.

From whatever source we procure light, it is the same in character, differing only in intensity. In its action upon matter, we have the phenomena of transparency, of reflection, of refraction, of colour, of polarization, and of vision, to engage our attention.

A beam of white light falls upon a plate of colourless glass, and it passes freely through it, losing but little of its intensity; that little being lost by reflection from the first surface upon which the light impinges. If the glass is roughened by grinding, we lose more light by reflection from the asperities of the roughened surface; but if we cover that face with any oleaginous fluid, as for instance turpentine, its transparency is restored. We have thus direct proof that transparency to light is due to molecular condition. This may be most strikingly shown by an interesting experiment of Sir David Brewster's: —

If a glass tube is filled with nitrous acid vapour, which is of a dull red colour, it admits freely the passage of the red and orange rays with some of the others, and, if held upright in the sunshine, casts a red shadow on the ground; by gently warming it with a spirit-lamp, whilst in this position, it acquires a much deeper and blacker colour, and becomes almost impervious to any of the rays of light; but upon cooling it again recovers its transparency.

It has also been stated by the same exact experimentalist, that having brought a purple glass to a red heat, its transparency was improved, so that it transmitted green, yellow, and red rays which it previously absorbed; but the glass recovered its absorptive powers as it cooled. A piece of yellowish-green glass lost its transparency almost entirely by being heated. Native yellow orpiment becomes blood-red upon being warmed, when nearly all but the red rays are absorbed; and pure phosphorus, which is of a pale yellow colour, and transmits freely all the coloured rays upon being melted, becomes very dark, and transmits no light.

Chemistry affords numerous examples of a very slight change

of condition, producing absolute opacity in fluids which were previously diaphanous bodies.(88)

Charcoal absorbs all the light which falls upon it, but in some of its states of combination, and in the diamond, it is highly transparent. In the same manner metals become transparent in their combinations; and gold and silver beaten into thin leaves are permeated by the green and blue rays. What becomes of the light which falls upon and is absorbed by bodies, is a question which we cannot yet, notwithstanding the extensive observations that have been made by some of the most gifted of men, answer in any way satisfactorily. In all probability it is permanently retained within their substances; and many of the experiments of exciting light in bodies when in perfect darkness, by the electric spark and other means, appear to support the idea of light becoming latent.

No body is perfectly transparent; some light is evidently lost in passing even through space, and still more in traversing our atmosphere.

Amongst the most curious instances of absorption is that which is uniformly discovered in the solar spectrum, if we examine it with a telescope. We then find that the coloured rays are crossed by a great number of dark bands, or lines, giving no light of any colour; these are generally called Fraunhofer's dark lines, as it was to the indefatigable exertions of that experimentalist, and by the aid of his beautiful instruments, that most of them were discovered and measured, and enumerated. It is quite clear that those lines represent rays which have been absorbed in their passage from the sun to the earth; although some of them have no doubt undergone absorption within the limits of the earth's atmosphere, we have every reason to believe, with Sir John Herschel, that the principal absorption takes place in the atmosphere of the sun.(89)

It has been shown by Dr. Miller, that the number of lines is continually varying with the alteration of atmospheric conditions;(90) and the evidences which have been afforded of peculiar states of absorption by the gaseous envelope of the earth, during the prosecution of investigations on the chemical agencies of the sun's rays, are of a sufficiently convincing character.

It has been calculated by Bouguer, that if our atmosphere, in its purest state, could be extended rather more than 700 miles from the earth's surface instead of nearly 40, as it is at present, the sun's rays could not penetrate it, and this globe would roll on in darkness and silence, without a vestige of vegetable form or of animal life. The same calculation supposes that sea-water loses all its transparency at the depth of 730 feet; but a dim twilight must prevail much deeper in the ocean.

The researches of Professor Edward Forbes have proved, that at the depth of 230 fathoms in the Ægean Sea, the few shelled animals that exist are colourless: no plants are found within that zone; and that industrious naturalist fixes the zero of animal life of those waters at about 300 fathoms.(91)

Our atmosphere, charged with aqueous vapour, serves, beyond the supply of oxygen it constantly affords for the support of life, to shield us from the intense action of the solar powers. By it we are protected from the destructive influences of the sun's light and heat, and enjoy those modified conditions which are most conducive to the healthful being of organic forms; and to it we owe "the blue sky bending over all," and those beauties of morning and evening twilight of which

——— Sound and motion own the potent sway,
Responding to the charm with its own mystery.

To defective transparency, or rather to variations of it, we must attribute, in part, the colours of permeable media. Thus, a glass or fluid appears yellow to the eye, because it has the property of admitting the permeation of a larger quantity of the yellow rays than of any others;—red, because the red rays pass it with the greatest freedom; and so on for every other colour. In most cases the *powers of transmission* and of reflection are similar; but it is not so in all; a variety of the Derbyshire fluor spar, and the precious opal, are striking instances to the contrary; and some glasses which transmit yellow light reflect blue; and a solution of quinine in water acidulated with tartaric or sulphuric acid, although perfectly transparent and colourless when held between the eye and the light, reflects, if viewed in a particular direction, a lively cerulean tint. These effects being due to the conditions of the surface, have been called epipolic phenomena. ([92]) There are some difficulties about the questions of transparency, which we shall see presently are not satisfactorily explained upon either of the received theories of light.

It is a general law of all the radiant forces, that whenever they fall upon any surface, a portion is thrown back or reflected at the same time as other portions are absorbed or transmitted. Upon this peculiarity appear to depend the phenomena of natural colour in bodies.

The white light of the sun is well known to be composed of several coloured rays. Or rather, according to the favorite theory, when the rate at which a ray undulates is altered, a different sensation is produced upon the optic nerve. The analytical examination of this question shows that to produce a red colour the ray of light must give 37,640 undulations in an inch, and 458,-000000,000000 in a second. Yellow light requires 44,000 undulations in an inch, and 535,000000,000000 in a second; whilst the effect of blue results from 51,110 undulations within an inch, and 622,000000,000000 in a second of time. Such results as these are among the highest refinements of science, and, when contrasted with the most sublime efforts of the imagination, appear immeasurably superior to them. ([53])

If a body sends back white light unchanged, it appears white; if the surface has the property of altering the vibration to that which is calculated to produce redness, the result is a red colour; the annihilation of the undulations produces blackness. By the other view, the beam of white light is supposed to consist of certain coloured rays, each of which has physical properties peculiar to itself, and thus is capable of producing different physiological effects. These rays falling upon a transparent or an opaque body suffer more or less absorption, and being thus dissevered, we have the effect of colour. A red body absorbs all the rays but the red; a blue surface, all but the blue; a yellow, all but the yellow; and a black surface absorbs the whole of the light which falls upon it.

That natural colours are the result of white light, and not innate properties of the bodies themselves, is most conclusively shown by placing coloured bodies in monochromatic light of another kind, when they will appear either of the colour of the light, or, by absorbing it, become black; whereas, when placed in light of their own character, the intensity of colour is greatly increasing.

Every surface has, therefore, a peculiar constitution, by which it gives rise to the diversified hues of nature. The rich and lively green, which so abundantly overspreads the surface of the earth, the varied colours of the flowers, and the numberless tints of animals, together with all those of the productions of the mineral kingdom, and of the artificial combinations of chemical manufacture, result from powers by which the relations of matter to light are rendered permanent, until its physical conditions undergo some change.

There is a remarkable correspondence between the geographical position of a region and the colours of its plants and animals. Within the tropics, where

"The sun shines for ever unchangeably bright,"

the darkest green prevails over the leaves of plants; the flowers and fruits are tinctured with colours of the deepest dye, whilst the plumage of the birds is of the most variegated description and of the richest hues. In the people also of these climes there is manifested a desire for the most striking colours, and their dresses have all a distinguishing character, not of shape merely, but of chromatic arrangements. In the temperate climates everything is of a more subdued variety: the flowers are less bright of hue; the prevailing tint of the winged tribes is a russet brown; and the dresses of the inhabitants of these regions are of a sombre character. In the colder portions of the earth there is but little colour; the flowers are generally white or yellow, and the animals exhibit no other contrast than that which white and black afford. A chromatic scale might be formed, its maximum point being at the equator, and its minimum at the poles.([34])

The influence of light on the colours of organized creation is well shown in the sea. Near the shores we find sea-weeds of the most beautiful tinctures, particularly on the rocks which are left dry by the tides; and the rich hues of the actiniæ, which inhabit shallow water, must have been often observed. The fishes which swim near the surface are also distinguished by the variety of their colours, whereas those which live at greater depths are gray, brown, or black. It has been found that after a certain depth, where the quantity of light is so reduced that a mere twilight prevails, the inhabitants of the ocean become nearly colourless That the sun's ray alone gives to plants the property of reflecting colour is proved by the process of blanching, or the etiolated state, produced by artificially excluding them from light.

By a triangular piece of glass, a prism, we are enabled to resolve light into its ultimate rays. The white pencil of light which falls on the first surface of the prism is bent from its path, and coloured bands of different colours are obtained. These bands or rays observe a curious constancy in their positions: the red ray is always the least bent out of the straight path: the yellow class comes next in the order of refrangibility; and the blue are the most diverted from the vertex of the prism. The largest amount of illuminating power exists in the yellow ray, and it diminishes towards either end.(95) It is not uninteresting to observe something like the same order of colour occurring at each end of the prismatic spectrum. The strict order in which the pure and mixed coloured rays present themselves is as follows:—

1. The *extreme red:* a ray which can only be discovered when the eye is protected from the glare of the other rays by a cobalt blue glass,—is of a crimson character—a mixture of the *red* and the *blue*, red predominating.(96)
2. The *red:* the first ray visible under ordinary circumstances.
3. The *orange:* red passing into and combining with yellow.
4. The *yellow:* the most intensely luminous of the rays.
5. The *green:* the yellow passing into and blending with the blue.
6. The *blue:* in which the light very rapidly diminishes.
7. The *indigo:* the dark intensity of blue.
8. The *violet:* the *blue* mingled again with the *red*—blue being in excess.
9. The *lavender gray:* a neutral tint, produced by the combination of the red, blue, and yellow rays, which is discovered most easily when the spectrum is thrown upon a sheet of tumeric paper.

Newton regarded the spectrum as consisting of seven colours of definite and unvarying refrangibility. Brewster and others appear to have detected a great diffusion of the colours over the spectrum, and regard white light as consisting only of three rays, which in the spectrum overlap each other; and from these — red, yellow, and blue — all the others can certainly be formed by combination in varying proportions. The truth will probably be found to be, that the ordinary prismatic spectrum is a compound of two spectra. We have already examined the heating power found in various parts of the spectrum, which, although shown to be in a remarkable manner in constant agreement with the colour of a particular ray, is not directly connected with it; that is, not as the effect of a cause, or the contrary. The chemical action of the solar rays, to which from its important bearings we shall devote a separate chapter, has, in like manner with heat, been confounded with the sun's luminous power; but although associated with light and heat, and modified by their presence, it must be distinguished from them.

We find the maximum of heat at one end of the spectrum, and that of chemical excitation at the other — luminous power observing a mean point between them. Without doubt we have all these powers acting reciprocally, modifying all the phenomena of each other, and thus giving rise to the difficulties which beset the inquirer on every side.

We have beautiful natural illustrations of luminous refraction in the rainbow and in the halo: in both cases the rays of light being separated by the refractive power of the falling rain or the moisture which constitutes a fog. In the simple toy of the child — the soap-bubble floating upon the air — the philosopher finds subjects for his contemplation; and from the unrivalled play of colours which he discovers in that attenuated film, he learns that the varying thicknesses of the surfaces influence, in a most remarkable manner, the colours of the sunbeam. Films of oil floating upon water present similar appearances; and the colours produced in tempering steel, are due entirely to the thickness of the oxidized surface produced by heat. The rich play of tints upon mother-of-pearl in the feathers of many birds, the rings seen in the cracks of rock crystal, or between the unequal faces of two pieces of glass, and produced by many chemical and indeed mechanical operations—are all owing to the same cause; that is, to the *interference of light*, or to rays proceeding from the same source, but crossing each other at very acute angles. If we take one of those steel ornaments which are formed by being covered with an immense number of fine lines, it will be evident that these striæ present many different angles of reflection, and that, consequently, the rays thrown back will, at some point or another, have a tendency to cross each other. The result of this is, that the quantity of light is augmented at some points of intersection, and annihilated at others.(97) Out of the investigation of the phenomena of diffraction, of the effects of thin and thick plates upon light, and the results of interference, has arisen the discovery of one of the most remarkable conditions within the range of physical science.

Two bright lights may be made to produce darkness.—If two pencils of light radiate from two spots very close to each other in such a manner that they cross each other at a given point, any object placed at that line of interference will be illuminated with the sum of the two luminous pencils. If we suppose those rays to move in waves, and the elevation of the wave to represent the maximum of luminous effect, then the two waves meeting, when they are both at the height of their undulation, will necessarily produce a spot of greater intensity. If now we so arrange the points of radiation, that the systems of luminous waves proceed irregularly, and that one arrives at the screen half an undulation before the other, the one in elevation falling into the depression of the other, a mutual annihilation is the consequence. This fact, paradoxical as it may appear, was broadly stated by Grimaldi, in the description of his experiments on the inflection of light, and has been observed by many others. The vibratory hypothesis, seizing upon the analogy presented by two systems of waves in water, explains this plausibly; but still upon examination it does not appear that the explanation is quite free from objection.([96])

Another theory, not altogether new to us, being indicated in Mayer's hypothesis of three primary colours (1775), and to be found as a problem in some of the Encyclopædias of the last century, has been put forth, in a very original manner, by that master-mind of intellectual Germany, Goëthe; and from the very comprehensive views which this poet-philosopher has taken of both animal and vegetable physiology (views which have been adopted by some of the first naturalists of Europe), we are bound to receive his theory of colours with every respect and attention.

Goëthe regards colour as the "thinning" of light; that is, by obstructing a portion of white light, yellow is produced; by reducing it still farther, red is supposed to result; and by yet farther retarding the free passage of the beam, we procure a blue colour, which is the next remove from blackness, or the absence of light. There is truth in this; it bears about it a simplicity which will satisfy many minds; by it many of the phenomena of colour may be explained: but it is insufficient for any interpretation of several of those recondite laws to which the other theories do give us some insight.

Newton may have allowed himself to be misled by the analogy presented between the seven rays of the spectrum and the notes in an octave. The mystic number, seven, may have clung like a fibre of the web of superstition to the cloak of the great philosopher; but the attack made by Goëthe upon the Newtonian philosophy betrays the melancholy fact of his being diseased with the lamentable weakness of too many exalted minds — an overweening self-esteem.

The polarization of light, as it has been unfortunately called — unfortunately, as conveying an idea of determinate and different points or poles, which only exists in theoretical analogy — presents to us a class of phenomena which promise to unclose the mysterious doors of the molecular constitution of bodies.

To give a familiar illustration of the distinction between ordinary and polarized light, we will suppose the use of a cylinder having a mirror at one end of it. If we point this to the sun, and receive the reflected image on a distant screen, we may turn the cylinder round on its axis, and the reflected ray will be found to revolve constantly and regularly with it. If, now, instead of receiving the ray direct from the sun, we allow a beam reflected from a glass plate at an angle of about 54° to fall upon the mirror, and then be reflected on to the screen, it will be found that the point of light has not the same properties as that previously examined; it is altered in its degree of intensity as the cylinder is turned round, has points of greatest brightness, and others at which it is lost in shadow. The polarized beam has been well compared to "a long, flat, straight stick," having *sides*, the ordinary ray being regarded as cylindrical. This remarkable change, as produced by the reflection of the ray from glass, was first observed by Malus, in 1808, ([90]) when amusing himself by looking at the beams of the setting sun, reflected from the windows of the Luxembourg Palace through a double-refracting prism. The same fact was, however, noticed, in the first instance, by Erasmus Bartolin, in Iceland-spar, a crystal, the primary form of which is a rhombohedron;

who perceived that the two images produced by this body were not in the same physical conditions.([100]) It was also studied by Huyghens and Sir Isaac Newton, and to our countryman is due the singular idea that a ray of light emerging from such a crystal has *sides*.

It must not be considered that this change in the character of the luminous beam is due to any of the powers of reflection or refraction of bodies; it is a property of matter independent of the other modes of action which it exercises over light.

The variety of striking effects produced by the polarization of light; the unexpected results which have sprung from the investigation of the laws by which it is regulated; and the singular beauty of many of its phenomena, have made it one of the most attractive subjects of modern science.

Ordinary light passes through transparent bodies, without producing any very striking effects in its passage; but it would appear that this thin band, this *extraordinary* beam of light, has the power of insinuating itself between the molecules of bodies, and by illuminating them, of enabling the eye to detect something of the structure of the mass. The chromatic phenomena of polarized light are so striking, that no description can convey an adequate idea of their character.

Spectra, more beautiful and intense than the prismatic image, systems of rings far excelling those of thin plates, and forms of the most symmetric order are constantly presenting themselves as the polarized ray is passed through various transparent substances. By altering the molecular arrangement of these bodies, either by heat or by pressure, a new order of phenomena at once present themselves, and by means of the polarized ray of light, differences in the chemical constitution of bodies, too slight to be discovered by any other mode of analysis, can be most readily and certainly detected.([101])

Although we cannot enter into any examination of all the conditions involved in the polarization of light, and the action of matter upon ordinary light, or when it is in this peculiar state, it will be readily conceived, from what has been already stated, that some most important properties are indicated, beyond those which science has made known.

What may be the especial use of light in this state of polarization in nature, it is at present difficult to determine; we are, however, certain that its agency must be necessary and most important, and we may hope that, through the industry of experimentalists, it will not be long before we add this knowledge to the stores already accumulated.

Every body, in some definite position, appears to have the power of producing this change upon the solar ray, as may be satisfactorily shown by examining any object with a polarizing apparatus.([102]) The sky at all times furnishes polarized light, which is most intense where it is blue and unclouded, and the point of maximum polarization is varied according to the relative position of the sun and the observer.([103]) It has been stated, that chemical change on the Daguerreotype plates and on photographic papers is more readily produced by the polarized than by the ordinary sunbeam.([104]) If this fact be established by future investigations, we advance a step towards the discovery so much desiderated of the part it plays in natural operations.

The refined and accurate investigations of Dr. Faraday stand prominently forward amid those which will redeem the present age from the charge of being superficial, and they will, through all time, be referred to as illustrious examples of the influence of a love of truth for truth's sake, in entire independence of the marketable value, which it has been unfortunately too much the fashion to regard. The searching examination made by this "interpreter of nature" into the phenomena of electricity in all its forms, has led him onward to trace what connexion, if any, existed between this great natural agent and the luminous principle.

By employing that subtile analyzer, a polarized ray, Dr. Faraday has been enabled to detect and exhibit effects of a most startling character. It has appeared to him that he has proved magnetism to have the power of influencing a ray of light in its passage through transparent bodies. A polarized ray is passed through a piece of glass or a crystal, or along the length of a tube filled with some transparent fluid, and the line of its path carefully observed; if, when this is done, the solid or fluid body is brought under powerful magnetic influence, such as we have at command by making a very energetic voltaic current circulate around a bar of soft iron, it will be found that the polarized light is disturbed; that, indeed, it does not permeate the medium along the same line.([105]) As this effect is most strikingly shown in bodies of the greatest density, and diminished in fluids, the particles of which are easily movable over each other, and has not hitherto been observed in any gaseous medium; the question has arisen,—does magnetism act directly upon the ray of light, or only indirectly, by producing a molecular change in the body through which the ray is passing? This question, so important in its bearings upon the connexion between the great physical powers, will, no doubt, before long receive a satisfactory reply.

Without any desire to generalize too hastily, we cannot but express a feeling, amounting to a certainty in our own mind, that those manifestations of luminous power, connected with the phenomena of terrestrial magnetism, which are so evident in all the circumstances attendant upon the exhibition of Aurora Borealis, and those luminous clouds which are often seen, independent of the Northern Lights, that a very intimate relation exists between the solar radiations and that power which so strangely gives polarity to this globe of ours.

In connexion with the mysterious subject of solar light, it is important that we should occupy a brief space in these pages with the phenomena of vision, which is directly dependent upon luminous radiation.

The human eye has been rightly called the "masterpiece of divine mechanism;" its structure is complicated, yet all the adjustments of its parts are as simple as they are perfect. The eyeball consists of four coats. The cornea is the transparent coat in front of the globe; it is the first optical surface, and this is attached to the sclerotic membrane, filling up the circular aperture in the white of the eye; the choroid coat is a very delicate membrane, lining the sclerotic, and covered with a perfectly black pigment on the inside; and close to this lies the most delicately reticulated membrane, the retina, which is, indeed, an extension of the optic nerve.

The eye, in its more superficial mechanical arrangements, presents exactly the same character as a camera obscura, the cornea being the lens which receives the images of objects and refracts them; but how infinitely more beautiful are all the arrangements of the organ of vision than the dark chamber of Baptista Porta!([105]) Arranged within this globe we have the aqueous humour, crystalline lens, and the vitreous humour: the first is a watery fluid, and the last a gelatinous one, while the crystalline lens is a little capsule of fluid membranaceous matter. These are for the purpose of correcting any aberrations of light, which are so evident in ordinary lenses, and giving to the whole an achromatic character, in which so perfect is everything in form and arrangement, that both spherical and chromatic aberrations are corrected, and by the agency of the cornea and the crystalline lens, perfect images are depicted on the retina, in a similar way to those very charming pictures which present themselves in the table of the camera obscura.

The seat of vision has been generally supposed to be the retina; but Mariotte has shown that the base of the optic nerve, which is immediately connected with the retina, is incapable of conveying an impression to the brain. The choroid coat, which lies immediately behind the retina, is regarded by Mariotte and Bernoulli as the more probable seat of vision. The retina, being transparent, offers no obstruction to the passage of the light onward to the black surface of the choroid coat, from which the vibrations are, in all probability, communicated to the retina and conveyed to the brain. Howbeit, upon one or the other of these delicate coats a distinct image is impressed by light, and the communication made with the brain possibly by a vibratory action. We may trace up the phenomena of vision to this point; we may conceive undulations of light, differing in velocity and length of wave, occasioning corresponding tremors in the neuralgic system of the eye; but how these vibrations are to communicate correct impressions of length, breadth, and thickness, no one has yet undertaken to explain.

It has, however, been justly said by Herschel: "It is the boast of science to have been able to trace so far the refined contrivances of this most admirable organ, not its shame to find something still concealed from scrutiny; for, however anatomists may differ on points of structure, or physiologists dispute on modes of action, there is that in what we *do* understand of the formation of the eye so similar, and yet so infinitely superior to a product of human ingenuity; such thought, such care, such refinement, such advantage taken of the properties of natural agents used as mere instruments, for accomplishing a given end, as force upon us a conviction of deliberate choice and premeditated design, more strongly, perhaps, than any single contrivance to be found whether in art or nature, and renders its study an object of the greatest interest."([107])

Analogy often is of great value in indicating the direction in which to seek for a truth; but analogical evidence, unless where the resemblance is very striking, should be received with caution. Mankind are so ready to leap to conclusions without the labour necessary for a faithful elucidation of the truth, that too often a few points of resemblance are seized upon, and an inference is drawn which is calculated to mislead.

There is a vague idea that the phenomena of sound bear a relation to those of light, — that there exists a faint resemblance between the chromatic and the diatonic scales. Sound, we know, is conveyed by the beating of material particles upon the auditory membrane of the ear, which have been set in motion by some distant disturbance of the medium through which it passes. Light has been supposed to act on the optic nerve in the same manner. If we imagine colour to be the result of vibrations of different velocities and lengths, we can understand that under some of these tremors, first established on the nerves, and through them conveyed to the brain, sensations of pain or pleasure may result, in the same way as sharp or subdued sounds are disagreeable or otherwise. Intensely coloured bodies do make an impression upon perfectly blind men; and those who, being born blind, know no condition of light or colour, will point out a difference between strongly illuminated red and yellow media. When the eyes are closed we are sensible to luminous influence, and even to differences of colour. We must consequently infer that light produces some peculiar action upon the system of nerves in general; this may or may not be independent of the chemical agency of the solar radiations; but certainly the excitement is not owing to any calorific influence. The system of nerves in the eye is more delicately organized, and of course peculiarly adapted to all the necessities of vision.

Thus far some analogy does appear to exist between light and sound; but the phenomena of the one are so much more refined than those of the other; — the impressions being, all of them, of a far more complicated character, that we must not be led too far by the analogical evidence in referring light, like sound, to mere material motion.

It was a beautiful idea that real impressions of external objects are made upon the seat of vision, and that they are viewed, as in a picture, by something behind the screen, — that these pictures become dormant, but are capable of being revived by the operations of the mind in peculiar conditions; but we can only regard it as a philosophical speculation of a high order, the truth or falsehood of which we are never likely to be enabled to establish. ([108])

That which sees will never itself be visible. The secret principle of sensation, — the mystery of the life that is in us, — will never be unfolded to finite minds.

Numerous experiments have been made from time to time on the influence of light upon animal life. It has been proved that the excitement of the solar rays is too great for the healthful growth of young animals; but, at the same time, it appears probable that the development of the functional organs of animals requires in some way, the influence of the solar rays. This might, indeed, have been inferred from the discovery that animal life ceases in situations from which light is absolutely excluded. The case of the Proteus of the Illyrian lakes may appear against this conclusion. This remarkable creature is found in the deep and dark recesses of the calcareous rocks of Adelsburg, at Sittich, and it is stated also in Sicily. Sir Humphry Davy describes the Proteus anguinus as " an animal to whom the presence of light is not essential, and who can live indifferently in air and in water, on the surface of the rock, or in the depths of the mud." The geological character of rocks, however, renders it extremely probable that these animals may have descended with the water, percolating through fissures from very near the surface of the ground. All the facts with which science has made us acquainted — and both natural and physical science has been labouring with most untiring industry in the pursuit of truth — go to prove that light is absolutely necessary to organization. It is possible, the influence of the solar radiations may extend beyond the powers of the human senses to detect luminous or thermic action, and that consequently a development of animal or vegetable forms may occur where the eye can detect no light; and under such conditions the Proteus may be produced in its cavernous abodes, and also those creatures which live buried deep in mud. Some further consideration of the probable agency of light will occupy us, when we come to examine the phenomena of vital forces.

Light is essentially necessary to vegetable life; and to it science refers the powers which the plant possesses of separating carbon from the air breathed by the leaves, and secreting it within its tissues for the purpose of adding to its woody structure. As, however, we have, in the growing plant, the action of several physical powers exerted to different ends at the same time, the remarkable facts which connect themselves with vegetable chemistry and physiology are deferred for a separate examination.

The power of the solar rays to produce in bodies that peculiar gleaming light which we call phosphorescence, and the curious conditions under which this phenomenon is sometimes apparent, independent of the sun's direct influence, present a very remarkable chapter in the science of luminous powers.

The phosphorescence of animals is amongst the most surprising of nature's phenomena, and to us is not the less so from our almost entire ignorance of the cause of it. Many very poetical fancies have been applied in description of these luminous creations; and imagination has found reasons why they should be gifted with these extraordinary powers. The glow-worm lights her lamp to lure her lover to her bower, and the luminous animalcules of the ocean are employed in lighting up the fathomless depths where the sun's rays cannot penetrate, to aid its monsters in their search for prey. "The lamp of love—the pharos—the telegraph of the night,—which scintillates and marks, in the silence of darkness, the spot appointed for the lover's rendezvous,"([100]) is but a pretty fiction; for the glow-worm shines in its infant state, in that of the larva, and when in its aurelian condition. Of the dark depths of the ocean it may be safely affirmed that no organized creation lives or moves in its gravelike silence to require this fairy-like aid. Fiction has frequently borrowed her creations from science. In these cases science appears to have made free with the rights of fiction.

The glow-worms (*lampyris noctiluca*), it is well known, have the power of emitting from their bodies a beautiful pale bluish-white light, shining during the hours of night in the hedgerow, like crystal spheres. It appears, from the observations of naturalists, that these insects never exhibit their light without some motion of the body or legs;—from this it would seem that the phosphorescence was dependent upon some nervous action, regulated at pleasure by the insect; for they certainly have the power of obscuring it entirely. If the glow-worm is crushed, and the hands or face are rubbed with it, luminous streaks, similar to those produced by phosphorus, appear. They shine with greatly-increased brilliancy in oxygen gas and in nitrous oxide. From these facts may we not infer that the process by which this luminosity is produced, whatever it may be, has a strong resemblance to that of respiration?

There are several varieties of flies, and three species of beetles of the genus *Elater*, which have the power of emitting luminous rays. The great lantern-fly of South America is one of the most brilliant, a single insect giving sufficient light to enable a person to read. In Surinam, a very numerous class of these insects are found, which often illuminate the air in a remarkable manner. In some of the bogs of Ireland a worm exists which gives out a bright green light; and there are many other kinds of creatures which, under certain circumstances, become luminous in the dark. This is always dependent upon vitality; for all these animals, when deprived of life, cease to shine.

At the same time we have many very curious instances of phosphorescence in dead animal and vegetable matter; the lobster among the Crustacea, and the whiting among fishes, are striking examples: decayed wood also emits much light under certain conditions of the atmosphere. This development of light does not appear to be at all dependent upon putrefaction; indeed, as this process progresses, the luminosity diminishes. We cannot but imagine that this light is owing, in the first place, to direct absorption by, and fixation within, the corpuscular structure of those bodies, and that it is developed by the decomposition of the particles under the influence of our oxygenous atmosphere.

The pale light emitted by phosphorus in the dark is well known; and this is evidently only a species of slow combustion, a combination of the phosphorus with the oxygen of the air. Where there is no oxygen, phosphorus will not shine; its combustion in chlorine or iodine vapour is a phenomenon of a totally different character from that which we are now considering. This phosphorescence of animal and vegetable matter has been regarded as something different from the slow combustion of phosphorus; but, upon examination, all the chemical conditions are found to be the same, and it is certainly due to a similar chemical change.

The luminous matter of the dead whiting or the mackerel may be separated by a solution of common salt or of sulphate of magnesia; by concentrating these solutions the light disappears; but it is again emitted when the fluid is diluted. The entire subject is, however, involved in the mystery of ignorance, although it is a matter quite within the scope of any industrious observer. The self-emitted light of the carbuncle of the romancer is realized in these remarkable phenomena.

The phosphorescence of some plants and flowers is not, perhaps, of the same order as that which belongs to either of the conditions we have been considering. It appears to be due rather to an absorption of light and its subsequent liberation. If a nasturtium is plucked during sunshine, and carried into a dark room, the eye, after it has reposed for a short time, will discover the flower by a light emitted from its leaves.

The following remarkable example, and an explanation of it by the poet Goëthe, is instructive:—

"On the 19th of June, 1799, late in the evening, when the twilight was deepening into a clear night, as I was walking up and down the garden with a friend, we very distinctly observed a flame-like appearance near the oriental poppy, the flowers of which are remarkable for their powerful red colour. We approached the place, and looked attentively at the flowers, but could perceive nothing further, till at last, by passing and repassing repeatedly, while we looked side-ways on them, we succeeded in renewing the appearance as often as we pleased. It proved to be a physiological phenomenon, and the apparent coruscation was nothing but the spectrum of the flower in the compensatory blue-green colour. The twilight accounts for the eye being in a perfect state of repose, and thus very susceptible, and the colour of the poppy is sufficiently powerful in the summer twilight of the longest days to act with full effect, and produce a compensatory image."(110)

The leaves of the *œnothera macrocarpa* are said to exhibit phosphoric light when the air is highly charged with electricity. The agarics of the olive-grounds at Montpelier have been observed to be luminous at night; but they exhibit no light, even in darkness, *during the day.* The subterranean passages of the coal mines, near Dresden, are illuminated by the phosphorescent light of the *rhizomorpha phosphoreus*, a peculiar fungus. On the leaves of the Pindoba palm, a species of agaric grows which is exceedingly luminous at night; and many varieties of the lichens, creeping along the roofs of caverns, lend to them an air of enchantment by the soft and clear light they diffuse. In a small cave near Falmouth, this luminous moss is very abundant; it is also found in the mines of Hesse; and according to Heinzmann, the *rhizomorpha subterranea* and *aidulæ* are also phosphorescent.

It is but lately that a plant, which abounds in the jungles in the Madura district of the East Indies, was sent to this country, which, although dead, was remarkably phosphorescent; and, when in the living state, the light which it emitted was extraordinarily vivid, illuminating the ground for some distance. Those remarkable effects may be due, in some cases, to the separation of phosphoretted hydrogen from decomposing matter, and, in others, to some peculiar electric manifestation.

The phosphorescence of the sea, or that condition called by fishermen *brimy*, when the surface, being struck by an oar, or the paddle-wheels of a steamer, gives out large quantities of light, has been attributed to the presence of myriads of minute insects which have the power of emitting light when irritated. The night-shining nereis *(Nereis noctiluca)* emits a light of great brilliancy, as do several kinds of the mollusca. The nereides attach themselves to the scales of fishes, and thus frequently render them exceedingly luminous. Some of the crustaceæ possess the same remarkable property;—twelve different species of *cancer* were taken up by the naturalists of the Zaire in the Gulf of Guinea.(111) The *cancer fulgens*, discovered by Sir Joseph Banks, is enabled to illuminate its whole body, and emits vivid flashes of light. Many of the medusæ also exhibit powerful phosphorescence.(112) These noctilucous creatures are, many of them, exceedingly minute, several thousands being found in a tea-cup of sea-water. They float near the surface in countless myriads, and when disturbed they give out brilliant scintillations, often leaving a train of light behind them.(113) By microscopic examination no other fact has been elicited than that these minute beings contain a fluid which, when squeezed out, leaves a line of light upon the surface of water. The appearance of these creatures is almost invariably on the eve of some change of weather, which would lead us to suppose that their luminous phenomena must be connected with electrical excitation; and of this, the investigations of Mr. C. Peach, of Fowey, communicated to the British Association at Birmingham, furnish the most satisfactory proofs we have as yet obtained.

Benvenuto Cellini gives a curious account of a carbuncle which would shine with great brilliancy in the dark. The same thing has been stated of the diamond; but it appears to be necessary, to procure these emissions of light, that the minerals should be first warmed near a fire. From this we infer that the luminous appearance is of a similar character to that of fluor spar, and of numerous other earthy minerals, which, when exposed to heat, phosphoresce with great brilliancy. Phosphorescent glow can also be excited in similar bodies by electricity, as was first pointed out by Father Beccaria, and confirmed by Mr. Pearsall.(114) These effects, it must be remembered, are distinct from the electric spark manifested upon breaking white sugar in the dark, or scratching sulphuret of zinc.

In the instances adduced there is not necessarily any exposure to the sunshine required. It is probable that two, if not three, distinct phenomena are concerned in the cases above quoted, and that all of them are distinct from animal phosphorescence, or the luminous appearance of vegetables. They, however, certainly prove, either that light is capable of becoming latent, or that it is only a condition of matter, in which it may be made manifest by any disturbance of the molecular forces. We have, in answer to this, very distinct evidence that some bodies are capable of deriving this property from the solar rays. Canton's phosphorus, which is a sulphuret of calcium, will, having been exposed to the sun, continue luminous for some time after it is carried into the dark; as will also the Bolognian stone, a sulphuret of barium. This result appears to be due to a particular class of the solar rays; for it has been found, if these sulphurets, spread smoothly on paper, are exposed to the influence of the solar spectrum for some little time, and then examined in the dark, that luminous spaces appear, exactly corresponding with the most refrangible rays, or those which excite chemical change; and one very remarkable fact must not be forgotten — the dark rays of the spectrum beyond the violet produce a lively phosphorescence, which is extinguished by the action of the rays of least refrangibility, or the heat-rays — whilst artificial heat, as a warm iron, produces a very considerable elevation of the phosphorescent effect. (115)

In these allied phenomena we have effects which are evidently dependent upon several dissimilar causes. The phosphorescence of the living animal is due, without doubt, to nervous excitation; that of the living vegetable to solar influence; and in the case of the mosses of caverns, &c. to that peculiar power which is connected with the chemical agency of the sun's rays, and which is now clearly proved to be capable of conduction. In the dead organic matter we have a purely chemical action developing the light, and in the inorganic bodies we have peculiar molecular constitution, by which an absorption of light appears to take place.

The subject is one of the greatest difficulty; the torch of science is too dim to enable us to see the causes at work in producing these marvellous effects. The investigation leads, to a certain extent, to the elucidation of many of the secrets of luminous action; and the determination of the question, whether light is an emanation from the sun, or only a subtile principle diffused through all matter, which is excited by solar influence, is intimately connected with the inquiry.

It has been stated that matter is necessary to the development of light; that no luminous effect would be produced if it were not for the presence of matter. Of this we not only have no proof, but such evidence as we have is against the position. There is no loss of light in the most perfect vacuum we can produce by any artificial means, which should be the case if matter was concerned in the phenomena of light as a cause.

Colour is certainly a property regulated by material bodies; or rather the presence of matter is necessary to the production of colour. Chlorine gas is a pale yellow, and nitrous vapour a yellowish red. These and one or two other vapours, which are near the point of condensation into fluids, are the only coloured gaseous or vaporiform bodies. The sky is blue, because the material particles of the atmosphere reflect back the blue rays. But we have more practical illustrations than this. The flame of hydrogen burning with oxygen gives scarcely any light; allow it to impinge on lime, a portion of which is carried off by the heat of the flame, and the most intense artificial light with which we are acquainted is produced. Hydrogen gas alone gives a flame in which nearly all but the blue rays are wanting: place a brush of steel or asbestos in it, and many of the other rays are at once produced. An Argand lamp, and more particularly the camphine Argand, gives a flame which emits most of the rays found in sunlight. Spirit of wine mixed with water, warmed and ignited, gives only yellow rays; add nitrate of strontian, and they become red; but nitrate of barytes being mixed with it, they are changed to green and yellow; salts of copper afford fine blue rays, and common salt, intense yellow ones. Many of these coloured rays and others can be produced in great power by the use of various solid bodies introduced into flame. This has not been sufficiently pointed out by authors; but it is clear from experiments that light requires the presence of matter to enable it to diffuse its glories. How is it that the oxygen and hydrogen flame gives so little light, and, with a solid body present, pours forth such a flood of brilliancy?

The production of artificial light by electrical and chemical agencies will necessarily find some consideration under their respective heads. There are numerous phenomena which connect themselves with luminous power, or appear to do so, which, in the present state of our knowledge, cannot come immediately within our attention. We are compelled to reserve our limited space for those branches of science which we are enabled to connect with the great natural operations constantly going on around us. Many of these more abstruse results will, however, receive some incidental notice when we come to examine the operation of the combined physical forces on matter.

We see in light a principle which, if it has not its source in the sun, is certainly dependent upon that luminary for its manifestations and powers. From that "fountain of light" we find this principle travelling to us at a speed which almost approaches the quickness of thought itself; yet by the refinements of science we have been enabled to measure its velocity with the utmost accuracy. The immortal poet of our own land and language, in his creation of Ariel, that "tricksy spirit," who could creep like music upon the waters, and girdle the earth in thirty minutes, appears to have approached to the highest point to which mere imagination could carry the human mind as to the powers of things ethereal. Science has, since then, shown to man that this "spirit, fine spirit," was a laggard in his tasks, and a gross piece of matter, when compared with the subtile essences which man, like a nobler Prospero, has now subdued to do him service. Light is necessary to life; the world was a dead chaos before its creation, and mute disorder would again be the consequence of its annihilation. Every charm which spreads itself over this rolling globe is directly dependent upon luminous power. Colours, and often, probably, forms, are the result of light, certainly the consequence of solar radiations. We know much of the mysterious influences of this great agent, but we know nothing of the principle itself. The solar beam has been tortured through prismatic glasses and natural crystals. Every chemical agent has been tried upon it, every electrical force in the most excited state brought to bear upon its operations, with a view to the discovery of the most refined of earthly agencies; but it has passed through every trial without revealing its secrets, and even the effects which it produces in its path are unexplained problems still, to tax the intellect of man.

Every animal and every plant is impelled to own that life and health are due to light; and even the crystallizing forms of inorganic matter, by bending towards it, confess its all-prevailing sway. From the sun to every planet revolving around that orb, and to the remotest stars which gleam through the vast immensity of heaven, we discover this power still in its brightness, giving beauty and order to these unnumbered creations, no less completely than to this small island of the universe. Through every form of matter we can mark its power, and from all we can, under certain conditions, evoke it in lustre and activity. Over all and through all light spreads its ethereal force, and manifests, in all its operations, powers which might well exalt the mind of Plato to the idea of an omniscient and omnipresent God. Science, with her Ithuriel wand, has, however, shown that light is itself an effect of a yet more exalted cause, which we can only refer to the source of every good and every perfect gift.

CHAPTER VIII.

ACTINISM — CHEMICAL RADIATIONS.

The Sun-ray and its Powers — Darkening of Horn Silver — Niepce's Discovery — Prismatic Spectrum — Refrangibility of Light, Heat, and Actinism — Daguerre's Discovery — Photography — Chemical Effects produced by Solar Radiations — Absorption of Actinism — Phenomena of the Daguerreotype — Chemical Change produced upon all Bodies — Power of Matter to restore its Condition — Light protects from Chemical Change — Photographs taken in Darkness — Chemical Effects of Light on Organized Forms — Chemical Effects of Solar Heat — Influence of Actinism on Electricity — Radiations in Darkness — Moser's Discoveries, &c.

Heat and light are derived from the sun, and we have attempted to show that not only are the phenomena of these two principles different, but that they can scarcely, in the present condition of our knowledge, be regarded as modified manifestations of *one* superior power. Associated with these two remarkable elements, others may exist in the solar rays. Electrical phenomena are certainly developed by both heat and light, and peculiar changes are produced by a short exposure to sunshine. Electricity may be merely *excited* by the solar rays, or it may flow like light from the sun. Chemical action may be only due to the disturbance of some diffused principle; or it may be directly owing to some agency which is radiated at once from the sun.

A sun-ray is a magical thing: we connect it in our fancy with the most ethereal of possible creations. Yet in its action on matter it produces colour; it separates the particles of solid masses farther from each other, and it breaks up some of the strongest forces of chemical affinity. To modern science is entirely due the knowledge we have gained of the marvellous powers of the sunbeam; and it has rendered us familiar with phenomena, to which the incantation-scenes of the Cornelius Agrippas of the Dark Ages were but ill-contrived delusions, and their magic mirrors poor instruments in comparison with the silver tablets of the photographic artist.

In the Dark Ages, or rather as the earliest gleams of the bright morning of industrious research were dispelling the mists of that phantom-peopled period, it was observed, for the first time, that the sun's rays turned a white compound black. Man must have witnessed, long before, that curious change which is constantly taking place in all vegetable colours: some darkening by exposure to sunlight, while others are bleached by the solar rays. Yet those phenomena excited no attention, and the world knew nothing of the mighty changes which were constantly taking place around them. The alchemists — sublime pictures of credulous humanity — toiling in the smoke of their secret laboratories, waiting and watching for every change which could be produced by fire, or by their "royal waters," caught the first faint ray of an opening truth; and their wild fancy, that light could change silver into gold, if they but succeeded in getting its subtile beams to interpenetrate the metal, was the clue afforded to the empirical philosopher to guide him through a more than Cretan labyrinth.([116])

The first fact recorded upon this point was, that horn silver blackened when exposed to the light. Without doubt many anxious thoughts were given by these alchemists to that fact. Here was, as it appeared, a mixing up of light and matter, and behold the striking change. It was a step towards the realization of their dreams. Alas, poor visionaries! in pursuing an ideality they lost the reality which was within their grasp.

Truths come slowly upon man, and long it is before these angel visits are acknowledged by humanity. The world clings to its errors, and avoids the truth, lest its light should betray their miserable follies.

At length a man of genius announced that "*No substance can be exposed to the sun's rays without undergoing a chemical change;*" but his words fell idly upon the ear. His friends looked upon his light-produced pictures as curious matters; they preserved them in their cabinets of curiosities: but the truth which he enunciated was soon forgotten. Howbeit, these words were recorded, and it is due to the solitary experimentalist of Châlons on the Saône, to couple the name of Niepce with the discovery of a fact which is scarcely second to the development of the great law of universal gravitation.([117]) But an examination awaits us, which, for its novelty, has more charms than most branches of science, and which, for the extensive views it opens to the inquirer, has an interest in nowise inferior to any other physical investigation.

The prismatic spectrum affords us the means of examining the conditions of the solar rays with great facility. In bending the ray of white light out of its path, by means of a triangular piece of glass, we divide it in a remarkable manner. We learn that heat is less refracted by the glass than the other powers; we find the maximum point of the calorific rays but slightly thrown out of the right line, which the solar pencil would have taken, had it not been interrupted by the prism; and the thermic action is found to diminish with much regularity on either side of this line. We discover that the luminous power is subject to greater refraction, and that its maximum lies considerably above that of heat; and that, in like manner, on each side the light diminishes, producing orange, red, and crimson colours below the maximum point, and green, blue, and violet above it. Again, we find that the radiations which produce chemical change are more refrangible than either of the others, and the maximum of this power is found at the point where light rapidly diminishes, and where scarcely any heat can be detected: it extends in full activity, above its maximum, to a considerable distance, where no trace of light can be discovered, and below that point, until light, appearing to act as an interfering agent, quenches its peculiar properties. These are strong evidences that light and actinism — as this principle has been named — are not identical: and we may separate them most easily and effectually from each other. Certain glasses, stained dark blue, with oxide of cobalt, admit scarcely any light; but they offer no interruption to the passage of actinism; on the contrary, a yellow glass, or a yellow fluid, which does not sensibly reduce the intensity of any one colour of the chromatic band of luminous rays, completely cuts off this chemical principle, whatever it may be. In addition to these, there are other results which we shall have to describe, which prove that, although associated in the solar beam, light and actinism are in constant antagonism.

When Daguerre first published his great discovery, the European public regarded his metal tablets with feelings of wonder; we have grown accustomed to the beautiful phenomena of this art, and we have become acquainted with a number of no less beautiful processes on paper, all of which, if studied aright, must convince the most superficial thinker, that a world of wonder lies a little beyond our knowledge, but within the reach of industrious and patient research. Photography is the name by which the art of sun-painting will be for ever known. We regard this as unfortunate, conveying as it does a false idea, — the pictures not being *light-drawn*. Could we adopt the name given by Niepce to the process, the difficulty would be avoided, since

Heliography involves no hypothesis, and strictly tells the undeniable fact, that our pictures are *sun-drawn.*

By whatever name we determine to convey our ideas of these phenomena, it is certain that they involve a series of effects which are of the highest interest to every lover of nature, and of the utmost importance to the artist and the amateur. By easy manipulation, we are now enabled to give permanence to the charming pictures which are produced by means of that pleasing invention of Baptista Porta, the *Camera Obscura.* Any image, which being refracted by the lens of this instrument falls upon the table in its dark chamber, may be secured with its most delicate gradations of shadows, upon either a metallic or a paper tablet.

Thus we are enabled to preserve the lineaments of those who have benefited their race by their genius or their bravery. By the agency of those very rays which give life and brilliancy to the laughing eye and roseate cheek, we can at once correctly trace the outline of the features we admire, and fill in those shadowy details which give the picture the charm of *vraisemblance.* The admirer of nature may copy her arrangements with strict fidelity. Every undulation of the landscape, every projecting rock or beetling tor, each sinuous river, and the spreading plains over which are scattered the homes of honest industry and domestic peace, intermingled with the towers or spires of those humble temples in which simple-hearted piety delights to kneel, — these, all of these, may, by the sunbeam which illuminates the whole, be faithfully pencilled upon our chemical preparations.

To the traveller how valuable is the process! The characteristic vegetation of distant lands, and the remains of hoar antiquity, speaking to the present of the past, and recording the histories of races which have fleeted away, may be alike secured to instruct "home-keeping wits," by the assistance of this beautiful art.

But it is necessary we name a few of the more striking phenomena of these changes. To commence with some of the more simple but no less important results.

Chlorine and hydrogen will not unite in darkness, nor will chlorine and carbonic oxide; but, if either of those gaseous mixtures are exposed to sunshine, they combine rapidly, and with explosion. A solution of the sulphate of iron in ordinary water may be preserved for a long time in the dark without undergoing any change; expose it to the sunshine, and a precipitation of oxide of iron is very rapidly produced. The mineral chameleon, the *manganesiate of potash* in solution, is almost instantly decomposed in daylight; but it is a long time before it undergoes any change in darkness. The same thing occurs with a combination of platinum and lime; and, indeed, it appears that precipitation is at all times, and under all circumstances, accelerated by the solar rays. As these precipitations are in exact agreement with the quantity of actinic radiation to which the solutions have been exposed, we may actually weigh off the relative quantities, representing in grains the equivalent numbers to the amount of actinism which has influenced the chemical compound.([118])

We have evidence which appears to prove that this agent may be absorbed by simple bodies, and that by this absorption an actual change of condition is produced, in many respects analogous to those allotropic changes which we have previously considered. Chlorine, in its ordinary state, and hydrogen, do not combine in the dark. If we employ the yellow medium of chlorine gas, for the purpose of analyzing the sun's rays previously to their falling upon some chemical compound which is sensitive to actinic power, we shall find that the chlorine obstructs all this actinism, and our unstable compound remains unchanged. But the chlorine gas which has interrupted this wonderful agent, appears to have absorbed it, and it is so far altered in its constitution that it will unite with hydrogen in the dark.(119) In like manner, if, of two portions of the same solution of sulphate of iron, one is kept in the dark and the other exposed to sunshine, it will be found that the solution which has been exposed will precipitate gold and silver from their combinations much more speedily than that which has been preserved in darkness.

The phenomena of the Daguerreotype involve many strange conditions. A plate of silver, on which a slight chemical action has been established by the use of iodine, is exposed to the lenticular image in the camera obscura. If allowed to remain under the influence of the radiations for a sufficient length of time, a faithful picture of the illuminated objects is delineated on the plate, as shown by the visible decomposition and darkening of the iodized surface. The plate is not, however, in practice allowed to assume this condition; after an exposure of a few seconds the radiant influence is cut off, and the eye cannot detect any evidence of change upon the yellow plate. It is now exposed to the vapour of mercury, and that metal in a state of exceedingly fine division is condensed upon the plate; but the condensation is not uniformly spread upon its face. The deposit of mercurial vapour is in exact proportion to the amount of chemical action produced. Is the change, by which this peculiar power of condensation is effected, a chemical, calorific, electrical, or merely a molecular one? The evidences, at present, are not sufficient to determine the question. In all probability we have the involved action of several forces. It is not necessary that a chemically prepared surface should be exposed to the sun's rays to exhibit this result. A polished plate of metal, of glass, of marble, or a piece of wood being partially exposed, will, when breathed upon, or presented to the action of mercurial vapour, show that a disturbance has been produced upon the portions which were illuminated, whereas no change can be detected upon the parts which were kept in the dark. It was thought, until lately, that a few chemical compounds, such as iodide of silver, the Daguerreotype and Calotype material, — chloride of silver, the ordinary photographic agent, — a few salts of gold, and one or two of lead and iron, were the only materials upon which these very remarkable changes were produced. We now know that it is impossible to expose any body, simple or compound, to the sun's rays, without its being influenced by this chemical and molecular disturbing power. To take our examples from inorganic nature, the granite rock which presents its uplifted head in firmness to the driving storm, the stones which genius has framed into forms of architectural beauty, or the metal which is intended to commemorate the great acts of man, and which in the human form proclaims the hero's deeds and the artists talent, are all alike destructively acted upon during the hours of sunshine, and, but for provisions of nature no less wonderful, would soon perish under the delicate touch of the most subtile of the agencies of the universe.

Niepce was the first to show that those bodies which underwent this change during daylight, possessed the power of restoring themselves to their original conditions during the hours of night, when this excitement was no longer influencing them. Resins, the Daguerreotype plate, the unprepared metal tablet, and numerous photographic preparations show this in a remarkable manner. ([120])

The picture which we receive to-night, unless we adopt some method of securing its permanency, fades away before the morning, and we try to restore it in vain. Upon some of our chemical preparations this is very remarkably shown, but by none in so striking a manner as by paper prepared with the iodide of platinum, which, being impressed with an image by heliographic power, restores itself in the dark, in a few minutes, to its former state of sensibility to sunshine. ([121]) The inference we alone can draw from all the evidences which the study of actino-chemistry affords, is, that the hours of darkness are as necessary to the inorganic creation as we know night and sleep are to the organic kingdom. But we must not forget that there does exist in the solar rays a balance of forces which materially modifies the amount of disturbing influence exerted by them on matter. Not only do we find that the chemical action is not extended over the whole length of the prismatic spectrum, but we discover that over spaces, which correspond with the maximum points of light and heat, a protective action is exerted. That is, that highly sensitive photographic agents, which blacken rapidly under exposure to diffused daylight, are entirely protected from change in full sunshine, if at the same time, as a strong light is thrown upon them by reflection, the yellow and extra red rays are brought to bear upon their surface. Not only so, but by employing media which will cut off the chemical rays of the spectrum, admitting only the luminous and calorific rays, we find that this power of protection is co-extensive with colour, or rather with *light*, and that a protected band, the length of the spectrum, remains white, whilst every other portion has blackened.([122])

Among the many curious instances of natural magic, none are more remarkable than an experiment not long since proposed, by which Daguerreotype pictures could be taken in absolute darkness. This is effected in the following manner: A large prismatic spectrum is thrown upon a lens fitted into one side of a dark chamber; and as we know that the actinic power resides in great activity beyond the violet ray, where there is no light, the only rays which we allow to pass the lens into the chamber, are those which are extra-spectral and non-luminous. These are directed upon any object, and from that object radiated upon a highly sensitive plate in a camera obscura. Thus a copy of the subject will be obtained by the agency of radiations which produce no sensible effect upon the optic nerve. This experiment is the converse of those which show us, that we may illuminate any object with the strongest sunlight which has passed such adio-actinic media — as yellow glass, the yellow solution of sulphuret of calcium, or of the bichromate of potash — and yet fail to secure any Daguerreotype copy of it, even upon the most exquisitely sensitive plate. Indeed, the image of the sun itself, when setting through an atmosphere which reduces its light to a red or rich yellow colour, not only produces no chemical change, but protects an iodized plate from it; and whilst every other part of the tablet gives a picture of surrounding objects in the ordinary character, the bright sun itself is represented by a spot upon which no change has taken place. ([123]) In tropical climes, where a brilliant sun is giving the utmost degree of illumination to all surrounding objects, all photographic preparations are acted upon more slowly than in the climate of England, where the light is less intense. As a remarkable instance of this fact, a circumstance may be mentioned, which is curiously illustrative of the power of light to interfere with actinism: —

A gentleman, well acquainted with the Daguerreotype process, took with him to the city of Mexico all the necessary apparatus and chemicals, expecting, under the bright light and cloudless skies of that climate, to produce pictures of superior excellence. Failure upon failure was the result; and although every care was used, and every precaution adopted, it was not until the rainy season set in that he could secure a good Daguerreotype of any of the buildings of that southern city.

Any reference to the chemical agency of LIGHT — *luminous rays* — has been avoided until we came to the consideration of this particular question of chemical change.

Upon organic compounds, as, for instance, upon the colouring matter of leaves and flowers, light does exert a chemical power;

and it is found that vegetable colours are bleached, not by rays of their own character, but by those which are *complementary* to them. A red dye fades under the influence of a green ray, and a yellow under that of a blue one, much more speedily than when exposed to rays of any other colour.([124]) It was long a question whether the decomposition of carbonic acid by plants was due to the luminous or the chemical rays. It is now clearly established that the luminous rays are the most active in producing this effect; which they do, in all probability, only indirectly, by exciting the vital powers of the organized structures, to which we would refer this phenomenon of gaseous decomposition.([125])

We have already noticed some chemical phenomena due to heat, particularly those experiments of Count Rumford's, which appeared to him to prove that the chemical agency of the sun's rays was due to its calorific power. A certain class of chemical phenomena, we know, may be produced by thermic action, both radiant and direct; but the only class of thermo-chemical action, which connects itself immediately with the solar radiations, belongs to a class of rays to which the name of Parathermic has been given, and to which belong the scorching, as it is called, of plants, the browning of the autumnal leaves, and probably the ripening of fruits.([126]) When we come to the consideration of those physical phenomena which belong to the growth of plants, all these peculiarities of solar action must be attended to in detail.

The manner in which we find the actinic power influencing electrical action, also shows us that this balance of force is continued through all the great principles of nature. If a galvanic arrangement is made, by which small quantities of metals may be slowly precipitated at one of the poles in the dark, and a similar arrangement be exposed to sunshine, it will be then found that no metal will be deposited: the sun's rays have interfered with the decomposing power of the electrical current. At the same time we learn, that, by throwing a beam of light upon a plate of copper, forming one of a galvanic pair, whilst it is under the influence of an acidulated solution, an additional excitation takes place, and the galvanometer will indicate the passage of an increased current of electricity. These two dissimilar actions appear enigmatical; but they may, there is no doubt, receive some solution from the influence of different rays on the contrary poles of the battery. One thing is quite evident, — electricity suffers a disturbance of one order, by light; and, by its associated principles in the sunbeam, an excitement of another. If a yellow glass is interposed between the galvanic arrangement and the sun, the electro-chemical precipitation goes on in the same manner as it would in darkness, and no extra excitement is produced upon the plates of the battery. From this it would appear that actinism and not light is to be regarded as the actively disturbing power. ([127])

We have already detailed many of the peculiarities of the different varieties of Phosphori, which would seem to be the result of light. Phosphorescence is, however, excited by those rays which produce no direct effect upon the eye. If we spread sulphuret of calcium upon paper, and expose it to the action of the solar spectrum, it is found to glow (in the dark) only over those spaces occupied by the violet rays and the dark rays beyond them, proving that the excitation necessary to the development of the phenomena of phosphorescence is due to a class of rays distinct from the true light-giving principle, and more nearly allied to that principle or power which sets up chemical decomposition.

Vision and colour, calorific action, chemical change, molecular disturbance, electrical phenomena, and phosphorescent excitation, all, each one with a strange duality, are connected with the sunbeam.

We find, when we receive solar spectra upon iodized plates, or on several kinds of photographic paper, that a line, over which no action takes place, is preserved at the top and bottom of the impressed image, and in many cases along the sides also. The only way in which this can be accounted for, as the spectrum represents the sun in a distorted form, is by supposing that rays come from the edges of the sun of a different character from those which proceed from the centre of that orb. (123)

May we not look upon the examples which have been given, as evidence of some such arrangement of the united principles as the following? Light proceeds with greatest power from the centre of the solar disc, although diffusing itself over every part; — as the force of light diminishes, calorific action is more strongly manifested, and a zone of heat surrounds the centre of light, beyond which, extending like an atmosphere to the sun himself, exists the mighty chemical power which we call actinism; and may not electricity be the result of the combined action of these three powers?

That actinism is one of the great powers of creation we have abundant proof. Nearly all the phenomena of change which have been referred to light, are now proved to be dependent upon actinic power; and beyond the influence which has been ascertained to be exerted by it upon all inorganic bodies, we shall have occasion to show still further the dependence of the vegetable and animal worlds upon its agency. The influence of the solar beams on vegetation is proved by common experience; the closer examination of its action on vegetable life is reserved for the chapter devoted to its phenomena. Of its influence on animals nothing is very correctly known; but some early experiments prove that they, like other organized bodies, are subject to all the radiant forces, as indeed, independent of experiment, our common experience teaches. Certain it is, that organization can take place only where the sun's rays can penetrate: where there is unchanging darkness, there we find all the silence of death. Prometheus stole fire from heaven, and gave the sacred gift to man, as the most useful to him of all things in his necessities: by the aid of it he could temper the severities of climate, render his food more digestible and agreeable, and illuminate the hours of darkness. So says the beautiful fiction of the Grecian mind,—which appears as the poetic dream or prophetic glance of a gifted race, who felt the mysterious truth they were yet unable to describe. Phaeton and Apollo are only other foreshadowings of the creative energies which dwell in the glorious centre of our universe. The poetry of the Hellenic people ascended above the littleness of merely human action, and sought to interpret the great truths of creation. Reflective, they could not but see that some mysterious powers were at work around them; imaginative, they gave to fine idealizations the government of those inexplicable phenomena. Modern science has shown what vastly important offices the solar rays execute, and that the principles discovered in a sunbeam are indeed the exciters of organic life, and the disposers of inorganic form.

It must not be forgotten that we have already alluded to a speculation which supposes this actinic influence to be diffused through all nature, to be indeed the element to which chemical force in all its forms is to be referred, and that it is merely excited by the solar rays. This hypothesis receives some support from the very peculiar manner in which chemical action once set up is carried on, independent of all extraneous excitement, after the first disturbance has been produced. If any of the salts of gold are exposed in connection with organic matter, as on paper, to sunshine for a moment, an action is begun, which goes on unceasingly in the dark, until the gold is reduced to its most simple state.(123) The same thing occurs with chromate of silver, some of the salts of mercury, argentine preparations mixed with protosulphate of iron or gallic acid, and some other chemical combinations. These progressive influences point to some law not yet discovered, which seems to link this radiant actinism with the chemical agent existing in matter.

This problem also connects itself with another class of facts which, although due in all probability, to a great extent, to calorific radiations, and hence known under the general term of Thermography, appear to involve both chemical and electrical excitation. From the investigations of Moser and of others we learn the very extraordinary fact, that even inanimate masses act and react upon each other by the influence of some *dark* radiations, and seem to exchange some of the peculiarities which they possess. This appears generally in the curious experiments which have been referred to, as confined merely to form or structure. Thus an engraved plate will give to a polished surface of metal or glass placed near it, after a very little time, a neat, distinct image of itself; that is, produce such a structural disturbance as will occasion the plate to receive vapour differently over those spaces opposite to the parts in cameo or in intaglio, from what it does over the opposite. If a piece of wood is used instead of a medal, there will, by similar treatment, be produced a true picture of the wood, even to the representation of its fibres.([130])

It has also been found that chemical decomposition is produced by the mere juxtaposition of different bodies. If iodide of gold or silver, perfectly pure, is placed upon a plate of glass, and a plate of copper covered with mercury is suspended over them, a gradual decomposition of those salts will take place, iodide of mercury will be formed, and the gold or silver salts will be reduced to a finely divided metallic state.([131])

A body, whose powers of radiating heat are low, being

brought near another whose radiating powers are more extensive, will, in the course of a short time, undergo such an amount of molecular disturbance, as will effect a complete change in the arrangement of its surface, and an impression of the body having the highest radiating powers will be made upon the other. This impression is dormant, but may be developed under the influence of vapour, or of oxidation.([132]) A body, such as charcoal, of low conducting power, being placed near another, such as copper, which is a good conductor, will, in a very short time, produce, in like manner, an impression of itself upon the metal plate. Thus any two bodies, whose conducting or radiating powers are dissimilar, being brought near each other, will occasion a molecular disturbance, or impress the one with the image of the other. However small the difference may be, an effect is perceived, and that of the most extraordinary kind, giving rise to the impression of actual images upon each surface exposed. It is thus that a print on paper may be copied on metal, by merely suspending it near a well-polished plate of silver or copper for a few days. The white and black lines radiate very differently; consequently, an effect is produced on the bright metal in the parts corresponding to the black lines, dissimilar to that which takes place opposite to the white portions of the paper; and, on the application of vapour, a true image of the one is found impressed upon the other.([133])

Bodies which are in different electrical states act upon each other in an analogous manner. Thus arsenic, which is highly electro-negative, will, when placed near a piece of electro-positive copper, readily impart to its surface an impression of itself; and so in like manner will other bodies, if in unlike conditions. Every substance physically different, (it signifies not whether as it regards colour, chemical composition, mechanical structure, calorific condition, or electrical state,) has a power of radiation by which a sensible change can be produced in a body differently constituted.

Fable has told us that the magicians of the East possessed mirrors in which they could at will produce images of the absent. Science now shows us that representations quite sufficient to deceive the credulous can be produced on the surface of polished metals without difficulty. A highly polished plate of steel may be impressed with images of any kind, which would remain invisible, the polished surface not being in the least degree affected, as it regards its reflecting powers; but that, by breathing over it, these dormant images would develop themselves, and fade away again as the condensed moisture evaporated from the surface. (134)

These, which are but a few of a series of results of as striking a character, serve to convince us that nature is unceasingly at work, that every atom is possessed of properties by which it influences every other atom in the universe, and that a most important class of natural phenomena appear, in the present state of our knowledge, to connect themselves directly with the radiant forces.

The alchemists observed that a change took place in chloride of silver exposed to sunshine. Wedgwood first took advantage of that discovery to copy pictures. Niepce pursued a physical investigation of the curious change, and found that all bodies were influenced by this principle radiated from the sun. Daguerre produced effects from the solar pencil which no artist could approach to; and Talbot and others extended the application. Herschel took up the inquiry; and he, with his usual power of inductive search and philosophical deductions, presented the world with a class of discoveries which show how vast a field of investigation is opening for the younger races of mankind, — a field in which a true spirit may reap the highest reward in the discovery of new facts, and to which we must look for a further development of those great powers with which we have already some slight acquaintance, and for the discovery of higher influences which are not yet dreamed of in our philosophy.

If music, with its mysteries of sound,
  Gives to the human heart a heavenward feeling;
The beauty and the grandeur which is found
Wrapping in lustre this fair earth around,
  Creation's wond'rous harmonies revealing,
  And to the soul in truth's strong tongue appealing,
With all the magic of those secret powers,
  Which, mingling with the lovely band of light,
The sun in constant undulation showers
To mould the crystals, and to shape the flowers,
  Or give to matter the immortal might
Of an embracing soul — should, from this sod,
Exalt our aspirations all to God.

CHAPTER IX.

ELECTRICITY.

Discovery of Electrical Force — Diffused through all Matter — What is Electricity? — Theories — Frictional Electricity — Conducting Power of Bodies — Hypothesis of two Fluids — Electrical Images — Galvanic Electricity — Effects on Animals — Chemistry of Galvanic Battery — Electricity of a Drop of Water — Electro-chemical Action — Electrical Currents — Thermo-Electricity — Animal Electricity — Gymnotus — Torpedo — Atmospheric Electricity — Lightning Conductors — Earth's Magnetism due to Electrical Currents — Influence on Vitality — Animal and Vegetable Developments — Terrestrial Currents — Electricity of Mineral Veins — Electrotype — Influence of Heat, Light, and Actinism on Electrical Phenomena.

If a piece of amber, *electrum*, is briskly rubbed, it acquires the property of attracting light bodies. This curious power excited the attention of Thales of Miletus; and from the investigations of this Grecian philosopher we must date our knowledge of one of the most important of the natural forces — Electricity.

If an inquiring mind had not been led to ask why does this curious natural production attract a feather, the present age, in all probability, would not have been in possession of the means by which it is enabled to transmit intelligence with a rapidity which is only excelled by that of the "swift-winged messengers of thought." To this age of application, a striking lesson does this amber teach. Modern utility would regard Thales as a madman. Holding a piece of yellow resin in his hand, rubbing it, and then picking up bits of down, or catching floating feathers, the old Greek would have appeared a very imbecile, and the *cui bono* generation would have laughed at his silly labours. But when he announced to his school that this amber held a soul, or essence, which was awakened by friction, and went forth from the body in which it previously lay dormant, and brought back the small particles floating around it, he gave to the world the first hint of a great truth which has advanced our knowledge of physical phenomena in a marvellous manner, and ministered to the refinements and to the necessities of civilization. Each phenomenon which presents itself to us, however simple it may appear to be, is an outward expression of some internal truth, the interpretation of which is only to be arrived at by assiduous study, but which, once discovered, directs the way to new knowledge, and gives to man a great increase of power.

Electricity appears to be diffused through all nature; and it is, beyond all doubt, one of the most important of the physical forces, in the great phenomena of creation. In the thunder-cloud, swelling with destruction, it resides, ready to launch its darts and shake the earth with its explosions: in the aërial undulations, silent and unseen, it passes, giving the necessary excitement to the organisms around which it floats. The rain-drop — the earth-girdling ocean — and the ringing waters of the hill-born river hold locked this mighty force. The solid rocks — the tenacious clays which rest upon them — the superficial soils — and the incoherent sands give us evidence of the presence of this agency; and in the organic world, whether animal or vegetable, the excitement of electrical force is always to be detected.

In the solar radiations, we have perhaps the prime mover of this power. In our atmosphere, when calm and cloudless, a great ocean of light, or when sombre with the mighty aspect of the dire tornado, we can constantly detect the struggle between the elements of matter to maintain an equilibrium of electrical force.

Diffused throughout matter, electricity is ever active; but it must be remembered that, although it is evidently a necessary agent in all the operations of nature, it is not the agent to which everything unknown is to be referred. Doubtless the influence of this force is more extensive than we have yet discovered; but that is an indolent philosophy which refers, without examination, every mysterious phenomenon to the influence of electricity.

The question, what is electricity? has ever perplexed, and still continues to agitate, the world of science. While one set of experimentalists have endeavoured to explain the phenomena they have witnessed, upon the theory that electricity is a peculiar subtile fluid pervading matter, and possessing singular powers of attraction and repulsion, another party find themselves compelled to regard the phenomena as giving evidence of the action of two fluids which are always in opposite states: while again, electricity has been considered by others as, like the attraction of gravitation, a mere property of matter.([135]) Certain it is that, in the manifestations of electrical phenomena, we have, as it appears, the evidence of two conditions of force; but in the states of *positive* or *negative*, of *vitreous* or *resinous* electricity, we have a familiar explanation in the assumption of some current flowing into or out of the material body, — of some principle which is ever active in maintaining its equilibrium, which, consequently, must act in two directions, and always exhibit that duality which is a striking characteristic of this subtile agent. It is a curious, and it should be an instructive fact, that each of the three theories of electricity is capable of proof, and has, indeed, been most ably supported by the rigorous analysis of mathematics. When we remember that some of the most enlightened investigators of this and the past age have severally maintained, in the most able manner, these dissimilar views, we should hesitate before we pronounce an opinion upon the cause or causes of the very complicated phenomena of electrical force.

Although we discover, in all the processes of nature, the manifestations of this principle of force in its characteristic conditions, it will be necessary, before we regard the great phenomena, to examine the known sources from which we can evoke the mighty power of electricity. If we rub a piece of glass or resin, we readily render this agent active; these substances appear, by this excitement, to become surrounded by an attractive or a repellent atmosphere. Let us rub a strip of writing paper with Indian rubber, or a strip of Gutta Percha with the fingers in the dark, and we have the manifestation of several curious phenomena. We have a peculiar attracting power; we have a luminous discharge in the shape of a spark; and we have very sensible evidence of muscular disturbance produced by applying the knuckle to the surface of the material. In each case we have the development of the same power.

Every substance in nature is an electric, and, if so disposed that its electricity may not fly off as it is developed, we may, by friction, manifest its presence, and, indeed, measure its quantity or its force. All bodies are not, however, equally good electrics; shell-lac, amber, resins, sulphur, and glass, exhibiting more powerfully the phenomena of frictional or mechanical electricity than the metals, charcoal, or plumbago. Solid bodies allow this peculiar principle to pass along them also in very different degrees. Thus electricity travels readily through copper and most other metals, platinum being the worst metallic conductor. It also passes through living animals and vegetables, smoke, vapour, rarefied air, and moist earth; but it is obstructed by resins and glass, paper when dry, oils, and dry metallic oxides, and in a very powerful manner by Gutta Percha.(130)

If, therefore, we place an electric upon any of those non-conducting bodies, the air around being well-dried, we are enabled to gather a large quantity of the force for the production of any particular effect. Taking advantage of this fact, arrangements are made for the accumulation and liberation at pleasure of any amount of electricity.

A Leyden phial, so called from its inventor, Muschenbroek, having resided at Leyden, is merely a glass bottle lined within and without, to within a few inches of the top, with a metal coating. If a wire or chain, carrying an electric current, is allowed to dip to the bottom of the bottle, the inner coat of the jar becomes charged, or gathers an excess, whilst the outer one is in its natural condition — one is said to be in a *positive*, and the other in a *negative* state. If the two coatings are now connected by a good conductor, as a piece of copper wire, passing from one to the other, the outside to the inside, a discharge, arising from the establishment of the equilibrium of the two coatings, takes place; and, if the connection is made through the medium of our bodies, we are sensible of a severe disturbance of the nervous system.

The cause of the conducting and non-conducting powers of bodies we know not; they bear some relation to their conducting powers for caloric; but they are not in obedience to the same laws. When we consider that resin, a comparatively soft body, in which, consequently, cohesive attraction is not very strong, is an imperfect conductor, and that copper, in which cohesion is much more powerful, is a good conductor, we may be disposed to consider that it is regulated by the closer approximation of the particles of matter. But in platinum the corpuscular arrangement must be much more dense than it is in copper, and yet it is, compared with it, a very bad conductor.([137])

We have now learned that we may, by friction, excite the electricity in a vitreous substance; but it must not be forgotten that we cannot increase the quantity which is, under ordinary conditions, natural to the electric; to do so, we must in some way establish a channel of communication with the earth, from which, through the medium we excite, we draw our supply. We have the means of confining this mighty force, and by placing conducting bodies upon insulating ones, we may retain a considerable quantity of it, in the same way as with the Leyden jar: but there is a constant effort to maintain a balance of conditions, and the body in which we have accumulated any extraordinary quantity soon returns to its natural state.

A very simple means may be adopted of showing what is thought to be one of the many evidences in favour of two electricities. If the wire carrying the current flowing from the machine, is passed over paper covered with nitrate of silver, it produces no change upon it; but if the wire which conveys the current to the instrument, when it is excited, is passed over the same paper, the silver salt is decomposed.([138]) We may, however, explain this result in a satisfactory manner, upon the hypothesis that the decomposition is produced by the abstraction or addition of electricity, rather than by any physical difference in the fluid itself. By this frictional electricity we may produce curious molecular disturbances, and give rise to re-arrangements, which have been called "electrical images," in glass, in stone, and in the apparently less tractable metals; — again, many of the great natural phenomena may be imitated by familiar arrangements of electrics.([139])

Voltaic electricity, as the active force produced by chemical change is commonly called, in honour of the illustrious Volta, is now to be considered. It differs from frictional electricity only in this: the electricity developed by friction of the glass plate or cylinder of the electrical machine, is a discharge with a sort of explosion, whereas that which is generated by chemical action is a steady flowing current. We may compare one to the ignition of a mass of gunpowder at once, and the other to the same quantity spread out into a very prolonged train.

There are numerous ways in which we may excite the phenomena of Voltaism, but in all of them the decomposition of one of the compounds employed appears to be necessary. This is the case in the arrangements of galvanic batteries, in which two dissimilar metals, zinc and copper, silver and platinum, or the like, are immersed in fluids; the zinc or the silver are gradually converted into soluble salts, which are dissolved, whilst the copper or platinum are protected from any action. The most simple manner of illustrating the development of this electricity is by placing a piece of silver on the tongue, and a piece of zinc or lead underneath it. No effect will be observed so long as the two metals are kept asunder, but when their hands are brought together, a slight sensation will pass through the tongue, and a saline taste be distinguished by the palate.

This, the germ of the most striking of the sciences, was noticed by Sulzer, fifty years before Galvani observed the convulsions in the limbs of frogs, when excited by the action of dissimilar metals; but the former paid little attention to the phenomenon, and the discovery led to no results.

When Galvani's observant mind was directed to the remarkable fact that the mere contact of two dissimilar metals with the moist surface of living muscles produced convulsions, there was an awakening in the soul of that philosopher to a great fundamental truth, which was nurtured by him, tried and tested, and preserved to work its marvels for future ages.

Although the world of science looks back to Volta as the man who gave the first true interpretation of this discovery, yet the ordinary world will never disconnect this important branch of physical science from the name of Galvani, and chemical electricity in all its forms will for ever be known under the familiar name of Galvanism.

Let us examine its phenomena, in its most simple phases: —

If we place a live flounder upon a piece of zinc, put a shilling on its back, and then touch both metals with the ends of the same metallic wire, the fish will exhibit painful convulsions. The zinc becomes oxidized by the separation of oxygen from the fluid on the surface, with which it is in contact, whilst hydrogen gas is liberated at the other metal. How the impulse which is derived from the zinc is transmitted through the body of the animal, or the tongue, to the silver or copper is the next consideration.

We can only understand this upon the supposition that a series of impulses are communicated in the most rapid manner along the connecting line; the idea of a current, although the term is commonly employed, tends to convey an imperfect impression to the mind. It would seem rather that a disturbance throughout the entire circuit is at once set up by a series of vibrations or impulses communicated from particle to particle, and along the strange network of nerves. One set of chemical elements have a tendency to develope themselves at that point where vibration is first communicated to the mass from a better conductor than it is, and another set at the point where it passes from the body to a better conductor than itself. The cause of this is to be sought for in the laws which regulate molecular constitution — by which chemical affinity is disturbed, — and a new attractive force exerted, in obedience to which the vital energy is itself agitated. We must not, however, forget that it is probable after all, although not yet susceptible of proof, that the electricity does nothing more than disturb or quicken the unknown principles upon which chemical and vital phenomena depend, being, indeed, a secondary agent.([140])

Notwithstanding our long acquaintance with the phenomena of Galvanism, there are but few who entertain a correct idea of the enormous amount of electricity which is necessary to the conditions of matter as we know it. To Faraday we are indebted for the first clear view of the most remarkable deduction of Science. He has proved, by a series of exceedingly beautiful and most conclusive experiments, that if the electrical power which holds a grain of water in combination, or which causes a grain of oxygen and hydrogen to unite in the right proportions to form water, could be collected and thrown into the condition of a Voltaic current, it would be exactly the quantity required to produce the decomposition of that grain of water, or the liberation of its elements, hydrogen and oxygen.([141])

By direct experiment it has been proved that one equivalent of zinc in a Voltaic arrangement evolves such a quantity of electricity in the form of a current, as, passing through water, will decompose exactly one equivalent of that fluid. The law has been thus expressed: The electricity which decomposes, and that which is evolved by the decomposition of a certain quantity of matter, are alike. The equivalent weights of bodies are those quantities of them which contain equal quantities of electricity; electricity determining the equivalent number, because it determines the combining force.([142])

The same elegant and correct experimentalist has shown that zinc and platinum wires, one-eighteenth of an inch in diameter, and about half an inch long, dipped into dilute sulphuric acid, so weak that it is not sensibly sour to the tongue, will evolve more electricity in one-twentieth of a minute than is given by thirty turns of a large and powerful plate electrical machine in full action, a quantity which, if passed through the head of a cat, is sufficient to kill it as by a flash of lightning. Pursuing this interesting inquiry yet further, it is found that a single grain of water contains as much electricity as could be accumulated in 800,000 Leyden jars, each requiring thirty turns of the large machine of the Royal Institution to charge it, — a quantity equal to that which is developed from a charged thunder-cloud. "Yet we have it under perfect command, — can evolve, direct, and employ it at pleasure; and when it has performed its full work of electrolization, it has only separated the elements of a single grain of water."

It has been argued by many that the realities of science will not admit of anything like a poetic view without degrading its high office; that poetry, being the imaginative side of nature, has nothing in common with the facts of experimental research, or with the philosophy which generalizes the discoveries of severe induction. If our science was perfect, and laid bare to our senses all the secrets of the inner world; if our philosophy was infallible, and always connected one fact with another through a long series up to the undoubted cause of all — then poetry, in the sense we now use the term, would have little business with the truth; it would, indeed, be lost or embodied, like the stars of heaven, in the brightness of a meridian sun. But to take our present fact as an example, how important a foundation does it offer upon which to build a series of thoughts, capable of lifting the human mind above the materialities by which it is surrounded, — of exalting each common nature by the refinement of its fresh ideas to a point higher in the scale of intelligence, — of quickening every impulse of the soul, — and of giving to mankind most holy longings.

What does science tell us of the drop of water? Two gases,

the one exciting life and quickening combustion, the other a highly inflammable air, are, by the influence of a combination of powers, brought into a liquid globe. We can, from this crystal sphere, evoke heat, light, electricity, and actinism in enormous quantities; and beyond these we can see powers or forces, for which, in the poverty of our ideas and our words, we have not names; and we learn that every one of these principles is engaged in maintaining the conditions of each drop of water which refreshes organic nature, and gives gladness to man's dwelling-place.

Has poetry a nobler theme than this? Agencies are seen like winged spirits of infinite power, each one working in its own peculiar way, and all to a common end, — to produce, under the guidance of omnipotent rule, the waters of the rivers and the seas. As the great ocean mirrors the bright heaven which overspreads it, and reflects back the sunlight and the sheen of the midnight stars in grandeur and loveliness; so every drop of water, viewed with the knowledge which science has given to us, sends back to the mind reflections of yet distant truths which, rightly followed, will lead us upwards and onwards in the tract of higher intelligences, —

> "To the abodes where the eternals are."

In the discoveries connected with electricity, we have results of a more tangible character than are as yet connected with the other physical forces; and it does appear that this science has advanced our knowledge of nature and of the mysteries of creation far more extensively than any other department of purely experimental inquiry.

The phenomena of electro-chemical action are so strange that we must return for a moment to the consideration of the decomposition of water, and the appearance of hydrogen at one pole, and of oxygen at the other. It appears that some confusion of our ideas has arisen from the views which have been received of the atomic constitution of bodies. We have been accustomed to regard water — to take that body as an example of all — as a compound of two gases, hydrogen and oxygen; an equivalent, or one atom of the first, united to an equivalent, or one atom of the last, forming one atom of water. This atom of water we regard as infinitely small; consequently, a drop of water is made up of many hundreds of these combined atoms, and a pint of water of not less than 10,000 drops. Now, if this pint of water is connected with the wires of a Galvanic battery, although these may be some inches apart, for every atom of oxygen liberated at one pole, an atom of hydrogen is set free at the other. It has been thought that an atom has undergone decomposition at one point, its oxygen being torn from it, and then there has arisen the difficulty of sending the atom of hydrogen through all the combined atoms of water across to the other pole. A series of decompositions and recompositions have been supposed to take place, and the communication of effects from particle to particle.

An attracting power for one class of bodies has been found in one pole, which is repellent to another class; and the reverse order has been detected at the opposite pole of a Galvanic arrangement.([143]) That is, the wire which carries the current from an excited zinc plate, has a relation to all bodies, which is directly opposite to that which is exhibited by the wire conveying the current from, or completing the circuit with, the copper plate. The one, for instance, collects and carries acids and the like, the other the metallic bases. At the extremity of one Galvanic wire, placed into a drop of water, oxygen is always liberated; and at the end of the other, necessary to complete the circuit with the battery, hydrogen is set free.

It appears necessary, to a clear understanding of what takes place in this experiment, that we should regard each mass, howsoever large, as the representative of a single atom. Nor is this difficult, as the following illustration will show:—

Let us take one particle of common salt (*chloride* of *sodium*) weighing less than a grain, and put it into a hundred thousand grains of distilled water. In a few minutes it has diffused itself through the whole of the fluid, and in every drop we can detect chlorine and soda. We cannot believe that this grain of salt has split itself up into a hundred thousand parts; we conceive rather that the phenomenon of solution is one of diffusion. One infinitely elastic body has interpenetrated with another.

Instead of an experiment with a pint of water, let us take our stand on Dover heights, and, with a gigantic battery at our command, place one wire into the ocean on our own shores, and convey the other through the air across the Channel, and let its extremity dip into the sea off Calais pier — the experiment is a practicable one — we have now an electrical circuit of which the British channel forms a part, and the result will be exactly the same as that which we may observe in a watch-glass with a drop of water.

We cannot suppose that the instantaneous and simultaneous effect, which takes place in the water at Calais and at Dover, is due to anything like what we have studied under the name of convection, when considering heat. We cannot conceive that the particle *A* excites the particle *B* next it, and so on through the series between the two shores; but regarding the Channel as one large drop, charged with the electric principle as we know it to be, it is excited by undulation or tremor throughout its width, and we have an equivalent of oxygen thrown off on one side of the line, and an exact equivalent of hydrogen at the other, the electro-chemical influence being exerted only where the current or motion is transferred from one medium to another.([144]) The imperfect character of this view is freely admitted; but regarding this remarkable class of phenomena as due to an internal excitement of the body under decomposition, rather than to any external one, no other, consistent with known facts, presents itself by which the effect can be explained. The fact stands as a truth; the hypothesis by which it is attempted to be interpreted is open to doubt, and it is opposed to some favourite theories.

Before we pass to the consideration of the other sources of electricity, it is important we should understand that no chemical or physical change, however slight it may be, can occur without the development of electrical power. If we dissolve a salt in water, if we mix two fluids together, if we condense a gas, or convert a fluid into vapour, electricity is disturbed, and may be made manifest to our senses.([145])

It has been shown that this power may be excited by friction (machine electricity); and it now remains to speak of the electricity developed by heat (thermo-electricity), and the electricity exhibited under nervous excitement by the gymnotus and torpedo (animal electricity); magnetism and its phenomena being reserved for a separate consideration.

If a bar of metal is warmed at one end and kept cool at the other, an electrical current circulates through the bar, and may be carried off by connection with any good conductor, and shown to exhibit the properties of ordinary electricity. The metals best suited for showing the effects of thermo-electricity appear to be bismuth and antimony. By binding two bars of these metals together at one end, and connecting the other ends with a Galvanometer, it will be discovered that an electric current passes off through the instrument by the slightest variation of temperature. Merely clasping the two metals, where bound together, with the finger and thumb, is sufficient to exhibit the phenomenon. By a series of such arrangements, which form what have been called thermo-electric multipliers, we obtain the most delicate measurers of heat with which philosophers are acquainted, and by the aid of which Melloni has been enabled to pursue his beautiful researches on radiant caloric.

That this electricity is identical with the other forms has been proved by employing the current thus excited for the purpose of producing chemical decomposition, magnetism, and electric light.(146)

The phenomenon of thermo-electricity — the discovery of Seebeck, is another proof of the very close connection of the physical forces. We witness their being resolved as it were into each other, electricity producing heat, and heat again electricity; and it is from these curious results that the arguments in favour of their intimate relations and actual identity have been drawn. It will, however, be found to be the best philosophy to regard these forces as dissimilar, until we are enabled to prove them to be only modified forms of one principle or power. At the same time it must not be forgotten that in natural operations we invariably find the combined action of several forces producing a single phenomenon. The important fact to be particularly regarded is, that we have evidence that every substance which is unequally heated, becomes the source of this very remarkable form of electricity.(147)

There exist a few fishes gifted with the very extraordinary power of producing electrical phenomena by an effort of muscular or nervous energy.

The *Gymnotus electricus*, or electrical eel, and the *Raia torpedo*, a species of ray, are the most remarkable. This power is, it would appear, given to these curious creatures for purposes of defence, and also for enabling them to secure their prey. The *Gymnotus* of the South American rivers, will, it is said, when in full vigour, send forth a discharge of electricity sufficiently powerful to knock down a man, or to stun a horse; while it can destroy fishes, through a considerable space, by exerting its strange artillery.([148])

Faraday's description of a *Gymnotus*, paralyzing and seizing its prey, is too graphic and important to be omitted.

"The *Gymnotus* can stun and kill fish which are in very various positions to its own body; but on one day, when I saw it eat, its action seemed to me to be peculiar. A live fish, about five inches in length, caught not half a minute before, was dropped into the tub. The *Gymnotus* instantly turned round in such a manner as to form a coil, inclosing the fish, the latter representing a diameter across it; a shock passed, and there, in an instant, was the fish struck motionless, as if by lightning, in the midst of the waters, its side floating to the light. The *Gymnotus* made a turn or two, to look for its prey, which, having found, he bolted, and then went about searching for more. A second smaller fish was given him, which, being hurt in the conveyance, showed but little signs of life, and this he swallowed at once, apparently without shocking it. The coiling of the *Gymnotus* round its prey had, in this case, every appearance of being intentional on its part, to increase the force of the shock, and the action is evidently well suited for that purpose, being in full accordance with the well-known laws of the discharge of currents in masses of conducting matter; and though the fish may not always put this artifice in practice, it is very probable he is aware of its advantages, and may resort to it in cases of need.([149])

Animal electricity has been proved to be of the same character as that derived from other sources. The shock and the spark are like those of the machine; and the current from the animal, circulating around soft iron, like Galvanic electricity, has the property of rendering it magnetic.

It is important that we should now review these conditions of electrical force in connexion with the great physical phenomena of nature.

It is sufficiently evident, from the results which have been examined, that all matter, whatever may be its form or condition, is for ever under the operation of the physical forces, in a state of disturbance. From the centre to the surface all is in an active condition: a state of mutation prevails with every created thing; and science clearly shows that influences are constantly in action which prevent the possibility of absolute repose.

Under the excitement of the several agencies of the solar beams, motion is given to all bodies by the circulation of caloric, and a full flow of electricity is sent around the earth to perform its wondrous works. The solar influences, which we regard as light, heat, actinism, and electricity, are active in effecting an actual change of state in matter, and in all probability in influencing the great magnetic phenomena of the world. The sunbeam of the morning falls on the solid earth, and its influence is felt to the very centre. The mountain-top catches the first ray of light, and its base, still wrapt in mists and darkness, is disturbed by the irradiating power. The crystalline gems, hidden in the darkness of the solid rock, are dependent for that form which makes them valued by the proud, on the influence of those radiations which they are one day to refract in beauty. The metals locked in the chasms of the rifted rocks are, for all their physical peculiarities, as dependent on solar influence as is the flower which lifts its head to the morning sun, or the bird which sings "at heaven's high gate."

Let us, then, examine how far electricity, as distinguished from the other powers, acts in producing any of these effects.

We find electricity in the atmosphere, in which it was first detected by the electrical kite of Dr. Franklin, and proved to be identical with that principle produced by the friction of glass. In the grandeur and terror of a thunder-storm many see nothing but manifestations of Almighty wrath. When the volleys of the bursting cloud are piercing the disturbed air, and the thunders of the discharge are pealing their dreadful notes above our heads, the chemical combinations of the noxious exhalations arising from the putrefying animal and vegetable masses of this earth are effected, and elements fitted for the purpose of health and vegetation are formed, and brought to the ground in the heavy rains which usually follow these storms. Science has taught man this — has shown him that the "partial evil" arising from the "winged bolt" is a "universal good;" and, more than this, it has armed him with the means of protecting his life and property from the influence of lightnings. By metallic rods, carried up a chimney, a tower, or a mast, we may form a channel through which the whole of the electricity of the most terrific thunder-cloud may be carried harmlessly into the earth or the sea; and it is pleasing to observe that at length prejudice has been overcome, and "conductors" are generally attached to high buildings, and to most of the ships of our navy.([150]) It was discovered that the devastating hailstorms of the south of France and Switzerland, so destructive to the vineyards and crops, were accompanied by evidences of great electrical excitation, and it was proposed to discharge the electricity from the air by means of pointed metallic rods. These have been adopted, and, it is said, with real advantage — each rod protecting an area of thirty or forty yards. Thus it is that science ministers to our service; and how much more pleasing is it to contemplate the lightning, with the philosopher, as an agent destroying the elements of pestilence, and restoring the healthfulness of the air we breathe, than with the romancer, to see in it only the dreaded aspect of a demon of destruction.

It has been thought, and much satisfactory evidence has been brought forward to support the idea, that the earth's magnetism is due to currents of electricity circulating around the globe; and the probability also derived from experiment is, that the great natural current is from east to west — that, indeed, it has an unvarying reference to the motion of the earth in relation to the sun.(151)

These terrestrial currents, as they have without doubt a very important bearing on the structural conditions of the rock-formations and the distribution of minerals, require an attentive consideration; but we must, in the first place, examine, as far as we know, the influences exerted, or supposed to be exerted, by atmospheric electricity.

The phenomena of vitality have, by many, been considered as immediately dependent upon its influence; and a rather extensive series of experiments might be quoted in support of this hypothesis. The researches of Philip on the action of the organs of digestion, when separated from their connection with the brain, but united with a Galvanic battery, have been proved by Dr. Reid to be delusive.(152) Since, as the organ is not removed from the influence of the living principle, it is quite evident that the electricity here is only secondary to some more important power. Matteucci has endeavoured to show that nervous action is due to electric excitation, and that electricity may be made a measurer of nervous irritability.(153) There can be no doubt that a peculiar susceptibility to excitement exists in some systems, and this is very strikingly shown in the disturbances produced by electric action; but in the experiments which have been brought forward we have only the evidence that a certain number of muscular contractions are exhibited in one animal by a current of electricity, giving a measured effect by the Voltameter, which are different from those produced upon another by a current of the same power. An attempt has recently been made by Mr. A. Smee to reduce the electrical phenomena connected with vitality to a more exact system than had hitherto been done. We cannot, however, regard the attempt as successful. The author has trusted almost entirely to analogical reasoning, which is, in science, always dangerous.([154]) In the development of electricity, from what may be particularly distinguished as the vital force, we see only the phenomena produced by the action of any two dissimilar chemical compounds upon each other. It has been thought that the structure of the brain presents an analogy to that of the galvanic battery, and the nerves represent the conducting wires. Although, however, some of the conditions appear similar, there are many which have no representatives in either the mechanical structure or the physical properties of the brain, so far as we know it. That the brain is the centre, the source, and termination of sensation is very clearly proved by physiological investigations. That the nerves are the media by which all sensation is conveyed to the brain, and also the instruments by which the will exerts its power over the muscles, is equally well established. But to say that we have any evidence to support the idea, that electricity has aught to do directly with these great physiological phenomena would be a bold assertion, betraying a want of due caution on the part of the investigator. That electric effects are developed during the operations of vitality is most certain. Such must be the case, from the chemical changes taking place during respiration and digestion, and the mechanical movements by which, even during external repose, the necessary functions of the body are carried on. Whether electricity is the cause of these, or an effect arising from them, we need not stop to examine, as this is, in the present state of our knowledge, a mere speculation. We have no evidence that electricity is an exciting power, but rather that it is one of those forces which tend to establish the equilibrium of matter. When disturbed — when its equilibrium is overset — it does, in its efforts to regain its stability, produce most remarkable effects. An electrical machine must be rubbed to exhibit any force. In all Galvanic arrangements, even the most simple, dissimilar bodies are brought together, and the latent electricity of both is disturbed; and, even in the magnet, it is only when this takes place, that its electrical powers are developed. In the *Gymnotus*, electricity appears to be dependent on the power of the will of the animal; but, even in this extraordinary fish, it is under only peculiar conditions that the electrical excitement takes place, and "what they inflict, they feel" during the restoration of that equilibrium which is necessary to their healthy state. In every case, therefore, we see that some power far superior to this is the ultimate cause; indeed, light and heat, and probably actinism, appear to stand superior to this principle; and on these, in some combined mode of action, in all probability, sensible electricity is dependent. Beyond even these elements, largely as they are engaged in the organic and inorganic changes of this world, there are occult powers which may never be understood by finite beings. We advance step by step from the most solid to the most ethereal of material creations, and we examine a series of extraordinary effects produced by powers which we know not whether to regard as material or immaterial, so subtile are they. On these, it appears, we may exhaust our inductive investigations — we may discover the laws by which these principles act upon the grosser elements, and develop phenomena of a very remarkable kind which have been unobserved or misunderstood. Whether light, heat, and electricity are modifications of one power, or different powers very closely united in action, is a problem we may possibly solve; but to know what they are, appears to be beyond the hopes of science; and it were idle to dream of elucidating the causes hidden beyond these forces, and by which they are regulated in all their actions on dead or living matter.

During changes in the electrical conditions of the earth and atmosphere, vegetables give indications of being in a peculiar manner influenced by this power. It is proved by experiments that the leaves of plants are among the best conductors of electricity, and it has hence been inferred that it must necessarily be advantageous to vegetation. That vegetable growth is, equally with animal growth, subject to electricity, as one of its quickening powers, must be admitted; but all experiments which have been fairly tried with the view of stimulating the growth of plants by its agency, have given results of a negative character.([155]) That a Galvanic arrangement may produce chemical changes in the soil, which may be advantageous to the plant, is probable; but that a plant can be brought to maturity sooner, or be made to develop itself more completely, under the direct action of electrical excitation, appears to be one of those dreams of science which will have a place amongst the marvels of alchemy and the fictions of astrology. An attentive examination of all the conditions necessary for the satisfactory development of the plant, will render it evident, that, although the ordinary electrical state of the earth and atmosphere must influence the processes of germination and vegetable growth, yet that any additional excitement must be destructive to them. The wonders wrought by electrical power are marvellous; a magic influence is exerted by it, and naturally the inquiring mind is led at first to believe that electricity is the all-powerful principle of creation; but a little reflection will serve to convince us that it is a suborbinate agent, although a powerful one.

In proceeding with our examination of the phenomena which

present themselves in connection with the terrestrial currents, we purposely separate magnetism from those more distinct electro-chemical agencies, which play so important a part in the great cosmical operations.

Electricity, we have already stated, flows through or involves all bodies; but, like heat, it appears to undergo a very remarkable change in becoming associated with some forms of matter. We have the phenomena of magnetism when an electric current circulates through a metallic wire, and it would appear that all other bodies acquire a peculiar polar condition under the influence of this principle, which will be explained in the next chapter.

The rocks, taken as masses, will not conduct an electric current when dry: granite, porphyry, slate, and limestone obstructing its passage even through the smallest spaces. But all the metallic formations admit of its circulating with great freedom. This fact it must, however, be remembered does not in any way interfere with the hypothesis of the existence of electricity in all bodies, in what we must regard as its latent state, from which, under prescribed conditions, it may be readily liberated. Neither does it affect the question of circulation, in relation to the great diffusion of electricity which we suppose to exist through all nature, and to move in obedience to some fixed law. We know that through the superficial strata electric currents circulate freely, whether they are composed of clay, sand, or any mixture of these with decomposed organic matter; indeed, that with any substance in a moist state they suffer no interruption.

The electricity of mineral veins has attracted much attention, and numerous investigations into the phenomena which these metalliferous formations present, have been made from time to time.([156])

By inserting into the mass of a copper *lode*, in situ, a metallic wire, which shall be connected with a measurer of Galvanic action, a wire also from the instrument being brought into contact with another *lode*, an immediate effect is generally produced, showing that a current is traversing through the wires from one *lode* to the other, and completing the circulation probably over the damp face of the rock in which the fissures forming the mineral veins exist.([157]) The currents thus detected are often sufficiently active to deflect a magnetic needle powerfully, to produce slowly electro-chemical decomposition, and to render a bar of iron magnetic. These currents must not, however, be confounded with the great electrical movement on which we had speculated. They are only to be detected in those mineral formations in which there is evidence of chemical action going on, and the greater the amount of this chemical operation, the more energetic are the electrical currents.([158]) We have, however, very good evidence that these local currents have, themselves, many peculiar influences. It not unfrequently happens that, owing to some great disturbance of the crust of the earth, a mineral vein is dislocated, and one part either sinks below, or is lifted above its original position; the fissures formed between the two being usually filled in with clay or some crystalline masses of more recent formation than the fissure itself. It is frequently found that these "*cross courses*," as they are called in mining language, contain ores of a different character from those which constitute the mineral vein; for instance, nickel, cobalt, and silver are not unfrequently discovered in them. When these metals are so found, they almost invariably occur between the ends of the dislocated lode, and often take a curvilinear direction, as if they were deposited along a line of electrical force.([159])

In the laboratory such an arrangement has been imitated, and in a mass of clay fixed between the Galvanic elements, after a short period, a distinct formation of a mineral vein has taken place.([160]) By the action, too, of weak electrical currents, Becquerel, Crosse, and others have been successful in imitating nature so far as to produce crystals of quartz and other minerals. In addition to this evidence, in support of the electrical theory of the origin of mineral veins, it can be experimentally shown that a schistose structure may be given to clays and sandstone by Voltaic action. ([161])

There is often a very remarkable regularity in the direction of mineral veins: throughout Cornwall, for instance, they most commonly have a bearing from the E. of N. to the W. of S. It has hence been inferred that they observe some relation to the magnetic poles of the earth. However this may be, it is certain that the ore in any lodes which are in a direction at right angles, or nearly so, to this main line, differs in character from that found in these, so called, east and west lodes ([162])

The sources of chemical action in the earth are numerous. Water percolating through the soil, and finding its way to great depths through fissures in the rocks, carries with it oxygen and various salts in solution. Water again rising from below, whether infiltrated from the ocean or derived from other sources, is usually of a high temperature, and it always contains a large quantity of saline matter. ([163]) By these causes alone chemical action must be set up. Chemical change cannot take place without a development of electricity: and it has been proved that the quantity of electricity required for the production of any change, is equal to that contained in the substances undergoing such change. Thus a constant activity is maintained within the caverns of the rock by the agency of the chemical and electrical elements, and mutations on a scale of great grandeur are constantly taking place under some directive force, probably magnetism.

The mysterious gnome, labouring — ever labouring — in the formation of metals, and the mischievous Cobalus of the mine, are the poor creations of superstition. A vague fear is spread amongst great masses of mankind relative to the condition of the dark recesses of the earth; a certain unacknowledged awe is experienced by many on entering a cavern, or descending a mine: not the natural fear arising from the peculiarity of the situation, but the result of a superstitious dread, the effect of a depraved education, by which they have been taught to refer everything a little beyond their immediate comprehension to supernatural causes. The spirit of demon worship, as well as that of hero worship, has passed from the early ages down to the present; and under its influence the genii of the East and the demons of the West have preserved their traditionary powers.

Fiction has employed itself with the utmost license in giving glowing pictures of treasures hidden in the earth's recesses. The caverns of Chilminar, the cave of Aladdin, the abodes of the spirits of the Hartz, and the dwellings of the fairies of England are gem-bespangled and gold-glistening vaults, to which man has never reached. The pictures are pleasing; but although they have the elements of poetry in them, and delight the young mind, they want the sterling test of scientific truth; and the wonderful researches of the plodding mineralogist have developed more beauty in the caverns of the dark rock than ever fancy painted in her happiest moments.

In all probability the action of the sun's rays upon the earth's surface, producing a constantly varying difference of temperature, and also the temperature which has been observed as existing at great depths, give rise to thermo-electrical currents, which may also play an important part in these results, which are thus briefly described.

In connexion with these great natural operations, explaining them, and being also, to some extent, explained by them, we have the very beautiful application of the electrotype.

Applying the views we have adopted to this beautiful discovery,([164]) the whole process by which these metallic deposits are produced will be yet more clearly understood. By the agency of the electric fluid, liberated in the Galvanic battery, a disturbance of the electricity of the solution of copper, silver, or gold, is produced, and the metal is deposited; but, instead of allowing the acid in combination to escape, it has presented to it some of the same metal as that revived, and, consequently, it combines with it, and this compound, being dissolved, maintains the strength of the solution.([165]) A system of revival is carried on at one pole, and one of abrasion, or more correctly speaking, of composition and solution, or change of state, at the other pole. By taking advantage of this very extraordinary power of electricity, we now form vessels for ornament or use, we gild or silver all kinds of utensils, and give the imperishability of metal to the most delicate productions of nature — her fruits, her flowers, and her insects; and over the finest labours of the loom we may throw coatings of gold or silver to add to their elegance and durability. Nor need we employ the somewhat complex arrangement of the battery: we may take the steel magnet, and, by mechanically disturbing the electricity it contains, we can produce a current through copper wires, which may be used, and is extensively employed, for gilding and silvering.([166]) The earth itself may be made the battery, and by connecting wires with its mineral bodies, currents of electricity have been collected, and those currents used for the production of electrotype deposit.([167])

The electrotype, as it has been called, is but one of the applications of electricity to the uses of man. This agent has recently been employed as the carrier of thought; and with infinite rapidity messages of importance, communications involving life, and intelligences outstripping the speed of coward crime, have been communicated by its means. There will be no difficulty in understanding the principle of this, although many of the nice mechanical arrangements, to insure precision, are of a somewhat elaborate character. The entire action depends on the deflection of a compass-needle by the passage of an electric current along its length. If at a given point we place a Galvanic battery, and at twenty or one hundred miles' distance from it there is fixed a compass-needle, between a wire brought from and another returning to it, the needle will remain true to its polar direction so long as the wires are free from the excited battery; but the moment connexion is made, the electricity of the whole extent of wire is disturbed, and the needle is thrown at right angles to the direction of the current. Provided a connexion between two points can be secured, however remote they are from each other, we thus, almost instantaneously, convey any intelligence. The effects of an electric current would appear at a distance of 576,000 miles in a second of time; and to that distance, and with that speed, is it possible, by Professor Wheatstone's beautiful arrangements, to convey whispers of love or messages of destruction.

The enchanted horse of the Arabian magician, the magic carpet of the German sorcerer, were poor contrivances, compared with the copper wires of the electrician, by which all the difficulties of time and the barriers of space appear to be overcome. In the Scandinavian mythology we find certain spiritual powers of evil possessed of the power of passing with imperceptible speed from one remote point to another, sowing the seeds of a common ruin amongst mankind. Such is the morbid creation of a wild yet highly-endowed imagination. The spirit of evil diffuses itself in a remarkable manner, and, indeed, we might almost assign to it the power of ubiquity; but in reality its advance is progressive, and time enters as an element into any calculation on its diffusion. May we not hope that the electrical telegraph, making, as it must do, the whole of the civilized world enter into a communion of thought, and, through thought, of feeling with each other, will bind us up in one common brotherhood, and that, instead of misunderstanding and of misinterpreting the desires and the designs of each other, we shall learn to know that such things as "natural enemies" do not exist? To hope to break down the great barrier of language is perhaps too much; but assuredly we may hope that, as we learn to know each other, as we must do when closer and more intimate relations are secured by the aids of science, the barrier of prejudice may be razed to the ground, and not one stone left to stand upon another? Our contentions, our sanguinary wars, consecrated to history by the baptism of blood, have in every, or in nearly every, instance sprung from the force of prejudice, or the mistakes of politicians, whose minds were narrowed to the limits of a convention formed for perpetuating the reign of ignorance.

And can anything be more in accordance with the spirit of all that we revere as holy, than the idea that the elements employed by the All Infinite in the works of physical creation shall be made, even in the hands of man, the ministering angels to the great moral redemption of the world? Associate the distant nations of the earth, and they will find some common ground on which they may unite. Mortality compels a dependence; and there are charities which spring up alike in the breast of the savage and the civilized man, which will not be controlled by the cold usages of pride, but which, like all truths, though in a still small voice, speak more forcibly to the heart than errors can, and serve as links in the great chain which must bind mankind in a common brotherhood. "None are all evil," and the best have much to learn of the amenities of life from him who yet lives in a "state of nature," or rather from him whose sensualities have prevailed over his intellectual powers, but who still preserves many of the noblest instincts, to give them no higher term, which other races, proud of their intelligence, have thrown aside. Time and space have hitherto prevented the accomplishment of this; electricity and mechanics promise to subdue both; and we have every reason to hope those powers are destined to accelerate the union of the vast human family.

Electrical power has also been employed for the purpose of measuring time, and by its means a great number of clocks can be kept in a state of uniform correctness, which no other arrangement can effect. A battery being united with the chief clock, which is itself connected by wires with any number of clocks arranged at a distance from each other, has the current continually and regularly interrupted by the beating of the pendulum, which interruption is experienced by all the clocks included in the electric circuit; and, in accordance with this breaking and making contact, the indicators or hands move over the dial with a constantly uniform rate. Instead of a battery, the earth itself has supplied the stream of electric fluid, with which the rate of its revolutions have been registered with the utmost fidelity.([168])

CHAPTER X.

MAGNETISM.

Magnetic Iron — Knowledge of, by the Ancients — Artificial Magnets — Electro-Magnets — Electro-Magnetism — Magneto-Electricity — Theories of Magnetism — The Magnetic Power of soft Iron and Steel — Influence of Heat on Magnetism — Terrestrial Magnetism — Declination of the Compass-needle — Variation of the Earth's Magnetism — Magnetic Poles — Hansteen's Speculations — Monthly and Diurnal Variation — Dip and Intensity — Thermo-Magnetism — Aurora Borealis — Magnetic Storms — Magnetic conditions of Matter — Dia-Magnetism, &c.

Agreeably with the view now generally received, that magnetism and electricity are but modifications of one force, since they are found to stand to each other in the relation of cause and effect, the separation which is here adopted, of the consideration of their several phenomena, may appear inappropriate. The importance, however, of all that is connected with magnetism, and the very decided difference which is presented by true magnetic action, and that of frictional or chemical electricity, is so great that it has been thought advantageous to adopt the present arrangement in reviewing the influences of terrestrial magnetism with which science has made us acquainted.

From a very early period a peculiar attractive force has been observed in some specimens of iron ore. Masses of this kind were found in Magnesia, and from that locality we derive the name given to iron in its polar condition. This is confirmed by the following lines by Lucretius : —

Quod superest agere incipiam, quo fœdere fiat
Natura lapis hic ut ferrum ducere possit,
Quem magnêta vocant patrio de nomine Graii
Magnêtum, quia sit patriis in finibus ortus.

Again we find Pliny employing the term *magnes*, to express this singular power. It was known to the ancients that the magnetic power of iron, and the electric property of amber, were not of the same character, but they were both alike regarded as miraculous. The Chinese and Arabians seem to have known it at a period long before that at which Europeans became acquainted with either the natural loadstone or the artificial magnet. Previously to A.D. 121, the magnet is distinctly mentioned in a Chinese dictionary; and in A.D. 419, it is stated in another of their books that ships were steered south by it. ([169])

The earliest popularly received account of its use in Europe is, that Vasco de Gama employed a compass in 1427, when that really adventurous navigator first explored the Indian seas. It is highly probable, however, that the knowledge of its important use was derived from some of the Oriental nations at a much earlier period.

We have some curious descriptions of the *leading stone* or loadstone, in the works of an Icelandic historian, who wrote in 1068. The mariner's compass is described in a French poem of the date of 1181; and from Torfæus's History of Norway, it appears to have been known to the northern nations certainly in 1266.

We have not to deal with the history of magnetic discovery; but so far as it tells of the strange properties which magnets are found to possess, and the application of this knowledge to the elucidation of effects occurring in nature.

A brown stone, in no respect presenting any thing by which it shall be distinguished from other rude stones around it, is found, upon close examination, to possess the power of drawing light particles of iron towards it. If this stone is placed upon a table, and iron filings are thrown lightly around it, we discover that these filings arrange themselves in symmetric curves, proceeding from some one point of the mass to some other; and upon examining into this, we shall find that the iron which has once clung to the one point, will be rejected by the other. If this stone is freely suspended, we shall learn also that it always comes to rest in a certain position, — this position being determined by these points, and some attractive force residing in the earth itself. These points we call its poles; and it is now established that this rude stone is but a weak representative of our planet. Both are magnetic: both are so in virtue of the circulation of currents of electricity, or of lines of magnetic force, as seen in the curves formed by the iron dust, and the north pole of the one attracts the south pole of the other, and the contrary. By a confusion of terms, we speak of the north pole of a compass-needle, meaning that point which is always opposite to the north pole of the earth: the truth being that the pole of the compass-needle, which is so forcibly drawn to the north, is a point in a contrary state, or, as we may express it, really a south pole.

There is a power of a peculiar kind, differing from gravitation, or any other attracting or aggregating force with which we are acquainted, which exists permanently in the magnetic iron stones, and also in the earth. What is this power?

Magnetism may be produced in any bar of steel, either by rubbing it with a loadstone, or by placing it in a certain position in relation to the magnetic currents of the earth, and, by a blow or any other means, disturbing its molecular arrangement. This principle appears to involve the iron as with an atmosphere, and to interpenetrate it. By one magnet we may induce magnetism in any number of iron bars without its losing any of its original force. As we have observed of the electrical forces already considered, the magnet constantly presents two points in which there is a difference manifested by the circumstance that they are always drawn with considerable power towards the north or south poles of the earth. That this power is of the same character as the electricity which we have been considering, is now most satisfactorily proved. By involving a bar of soft iron which, being without any magnetic power, is incapable of sustaining even an ounce weight, with a coil of copper wire, through which a galvanic current is passing, the bar will receive, by induction from the current, an enormous accession of power, and will, so long as the current flows around it, sustain many hundred pounds weight, which, the moment the current is checked, fall away from it in obedience to the law of gravity. Thus the mere flow of this invisible agent around a mass of metal possessing no magneto-attractive power, at once imparts this life-like influence to it, and as long as the current is maintained, the iron is endowed with this surprising energy.

This discovery, which we owe to the genius of Oersted, and which has, indeed, given rise to a new science, electro-magnetism, may be regarded as one of the most important additions made to our knowledge.

Current electricity is magnetic; iron is not necessary to the production of magnetic phenomena, although by its presence we secure a greater amount of power. The copper wires which complete the circuit of a galvanic battery will attract and hold up large quantities of iron filings, and the wires of the electric telegraph will do the same, while any signal is being passed along the line. Again, all the phenomena common to galvanic electricity can be produced by merely disturbing the power permanently secured in the ordinary magnet. It was thought that magnets would become weakened by this constant disturbance of their magnetism; but, since its application to the purposes of manufacture, and magneto-electricity has been employed in electro-plating, it has been found that continued action for more than a year, during which enormous quantities of electricity have been thus given out and employed in producing chemical decomposition, has not, in the slightest degree, altered their powers.

Thus a small bar of metal is shown to be capable of pouring out, for any number of years, the principle upon which the phenomena of magnetism depends.

There are, however, differences, and striking ones, between ordinary and magnetic electricity. In the magnet, we have a power at rest, and in the electrical machine or galvanic battery, a power in motion. Ordinary electricity is stopped in its passage by a plate of glass, of resin, and many other substances; but magnetism passes these with freedom, and influences magnetic bodies placed on the other side. It would appear, though we cannot explain how, that magnetism is due to some lateral influence of the electric currents. A magnetic bar is placed over a copper wire, and it hangs steadily in the direction of its length; an electric current is passed along it, and the magnet is at once driven to place itself across the wire. Upon this experiment, in the main, Ampère founds his theory of terrestrial magnetism. He supposes electrical currents to be traversing our globe from east to west, and thus, that the needle takes its direction, not from the terrestrial action of any fixed magnetic poles, but from the repulsion of these currents, as is the case with the wire.

It has been found that wires, freely suspended, along which currents were passing in opposite directions, revolve about each other, or have an inclination to place themselves at right angles, thus exhibiting the same phenomenon as the magnet and the conducting wire. So far, the hypothesis of Ampère leads us most satisfactorily. We see in the magnet one form of electricity, and in the machine or battery another. But why should not the electricity of the magnet, electricity at rest, exhibit the same powers as this force in motion?

Oersted, whose theory led him to discovery of the fact of electro-magnetic action, regards the phenomena of a current passing a wire, and its action on a needle, as evidence of two fluids, positive and negative, traversing in opposite directions, and mutually attracting and repelling, so that, indeed, they pass the wires in a series of spirals; that in the magnet, by some peculiar property of the iron, this conflict of the currents is reduced to an equilibrium, and its power becomes manifested in its attractive force. ([170]) It is a curious fact that iron becomes magnetic in a superior degree to any other metal; that steel permanently retains any magnetism imparted to it; but that soft iron rapidly loses its magnetic power. This must be in virtue of some peculiar arrangement of the molecules, or some unknown physical condition of the atoms of the mass, by which a continued influence is retained by the steel probably in a state of constant internal circulation. It has, however, been shown that soft iron, under certain circumstances, may be made to retain a large amount of magnetic force. ([171])

If a horse-shoe shaped bar of soft iron is rendered magnetic by the circulation of an electric current around it, while its two ends are united by an armature, of soft iron also, so that while the current is passing it is capable of supporting many hundred pounds weight; it will be found that a considerable weight may be supported when the current is stopped, provided the armature is carefully kept in contact, and it will retain this power for many years; but remove the connecting piece of iron and the bar immediately loses all its magnetism, and will not support even the armature itself. This fact appears to confirm the idea that magnetism is due to the retention of electricity, and that steel possesses the property of equalizing the opposing forces, or of binding this principle to itself like an atmosphere.

The influence of heat on magnetism is so remarkable a proof of the dependence of this power upon molecular arrangement, that it must not escape our notice. To select but one of many experiments by Mr. Barlow, it was found that in a bar of malleable iron, in which the magnetic effect when cold was + 30° 0′, all polarity ceased at a white heat, that it was scarcely appreciable at a red heat, but that at a blood-red heat it was equal to + 41° 0′.(172)

The more closely we examine the peculiarities of the magnetic power, and particularly as they are presented to us in its terrestrial action, the more surprising will its influences appear to be. We have discovered a natural cause which certainly exercises a very remarkable power over matter, and we have advanced so far in our investigations as to have learnt the secret of converting one form of force into another, or of giving to a principle, produced by one agency, a new character under new conditions; of changing, in fact, electricity into magnetism, and from magnetism again evolving many of the effects of electrical currents.

If a magnetic bar is freely suspended above the earth, it takes, in virtue of some terrestrial power, a given direction, which is an indication of the earth's magnetic force. Whether this is the consequence of the currents of electricity, which Ampère supposes to circulate around the globe, from east to west, or the result of points of attraction in the earth itself, the phenomenon is equally wonderful. To whatever cause we may refer the visible effects, it appears certain that this earth is composed of particles in a magnetic state, the character varying with physical conditions, and that terrestrial magnetic force is the collective action of all the atoms of this planetary mass.(173)

The constancy with which a magnetized needle points to a certain spot, renders it one of the most important instruments to the practical and the scientific man. The wanderer of the ocean or of the desert is enabled, without fear of error, to pursue his path, and in unknown regions to determine the azimuth of objects. The miner or the surveyor finds in the magnetic compass the surest guide in his labours, and the experimentalist is for ever studying its indications

"True as the needle to the pole,"

has passed into a proverb among mankind, but the searching inquiry of modern observers has shown that the expression is correct only with certain limitations. There are but two lines on the surface of the earth on which the needle points true north, or where the magnetic and the geographical north correspond. These are called lines of *no variation*, or, as they have also been designated, *agonic lines*, and one is found in the eastern and the other in the western hemisphere. The American line is singularly regular, passing in a south-east direction from latitude 60° to the west of Hudson's Bay, across the American lakes, till it reaches the South Atlantic ocean, and cuts the meridian of Greenwich in about 65° south latitude. The Asiatic line of *no variation* is very irregular, owing, without doubt, to local interferences; it begins below New Holland, in latitude 60° south, it bends westward across the Indian ocean, and from Bombay has an inflection eastward through China, and then northward across the sea of Japan, till it reaches the latitude of 71° north, when it descends again southward, with an immense semicircular bend, which terminates in the White Sea.

Hansteen has thought that there are two points in each hemisphere which may be regarded as stronger and weaker poles on opposite sides of the poles of revolution. These are called the magnetic poles of the earth, or by Hansteen *magnetic points of convergence.* These four points are considered to have a regular motion round the globe, the two northern ones from west to east, and the two southern ones from east to west. By the assistance of recorded observations, Hansteen has calculated the periods of these revolutions to be as follows:—

The weakest north pole in 860 years.
The strongest north pole in 1746 years.
The weakest south pole in 1304 years.
The strongest south pole in 4609 years.

There are some points of speculation on which Hansteen has ventured which have been smiled at as fanciful; but they may rather indicate an amount of knowledge in the Brahminical and Egyptian priesthood, beyond what we are usually disposed to allow them, and prove that their observations of nature had led to an appreciation of some of the most remarkable harmonies of this mysterious creation.

The above terms are exceedingly near 864, 1246, 1728, 4320, and those numbers are equal to the mystic number of the Indians, Greeks, and Egyptians, 432 multiplied by 2, 3, 4, and 10. On these the ancients believed a certain combination of natural events to depend, and, according to Brahminical mythology, the duration of the world is divided into four periods, each of 432,000 years. Again, the sun's mean distance from the earth is 216 radii of the sun, and the moon's mean distance 216 radii of the moon, each the half of 432. Proceeding with this very curious examination, Hansteen says, 60 multiplied by 432 equals 25,920, the smallest number divisible at once by all the four periods of magnetic revolution, and hence the shortest time in which the four poles can complete a cycle, and return to their present state, and *which coincides exactly with the period in which the precession of the equinoxes will amount to a complete circle,* reckoning the precession at a degree in seventy-two years.([174])

When we consider the phenomena of terrestrial magnetism carefully, it appears to indicate the action of a power external to the earth itself, and, as Hansteen conceives, having its origin from the action of the sun, heating, illuminating, and producing a magnetic tension, in the same manner as it produces electrical excitation and actino-chemical action.

The movements of these magnetic poles have been the subject of extensive and most accurate observation in every quarter of the globe. In London, during 1657–1662, there was no magnetic variation; the agonic line passing through it. The variation steadily increased, until, in 1815, it amounted to 24° 15′ 17″, since which time it has been slowly diminishing. In addition to this great variation, we have a regular annual change dependent on the position of the sun, in reference to the equinoctial and solstitial points, which has been discovered by Cassini, and investigated by Arago and others. Also a diurnal variation, which movement appears to commence early in the morning, moving eastward until half-past seven, A. M., when it begins to move westward until two, P. M., when it again returns to the east, and in the course of the night reaches the point from which it started twenty-four hours before.

We have also remarkable variations in what is termed the *dip* of the needle. It is well known that a piece of unmagnetized steel, if carefully suspended by its centre, will swing in a perfectly horizontal position, but, if we magnetize this bar, it will immediately be drawn downwards at one end. The force of the earth's polarity, attracting the dissimilar pole, has caused it to *dip*.

There is, in the neighbourhood of the earth's equator, and cutting it at four points, an irregular curve, called the magnetic equator, or *aclinic* line, where the needle balances itself horizontally. As we proceed from this line towards either pole the dip increases, until, at the north and south poles, the needle takes a vertical position. The *intensity* of the earth's magnetism is also found to vary with the position, and to increase in a proportion which corresponds very closely with the dip. But the *intensity* is not a function of the dip, and the lines of equal intensity *isodynamic lines*, are not parallel to those of equal dip. We have already remarked on the diurnal variation of the declination of the needle; we know, also, that there exists a regular monthly and daily change in the magnetic intensity. The greatest monthly change appears when the earth is in its perihelion and aphelion, in the months of December and June,—a maximum then occurs; and about the time of the equinoxes, when our planet is at the greatest mean distance from the sun, a minimum is detected.([175])

The daily variation of intensity is greatest in the summer, and least in the winter. The magnetism is generally found to be at a minimum when the sun is near the meridian; its intensity increasing until about six o'clock, when it again diminishes.([176])

What striking evidence all these well-ascertained facts give of the dependence of terrestrial magnetism on solar influence; and in further confirmation of this view, we find a very remarkable coincidence between the lines of equal temperature — the isothermal lines, and those of equal dip and magnetic intensity.

Sir David Brewster first pointed out that there were, in the northern hemisphere, two poles of maximum cold; these poles agree with the magnetic points of convergence; and the line of maximum heat, which does not run parallel to the earth's equator, is nearly coincident with that of magnetic power. Since Seebeck has shown us that electrical and magneto-electrical phenomena can be produced by the action of heat upon metallic bars, we have, perhaps, approached towards some faint appreciation of the manner in which the solar calorific radiations may, acting on the surface of our planet, produce electrical and magnetic effects. If we suppose that the sun produces a disturbance of the earth's electricity along any given line, in all directions at right angles to that line, we shall have magnetic polarity induced.(177) That such a disturbance is regularly produced every time the sun rises has been sufficiently proved by many observers.

In 1750, Wargentin noticed that a very remarkable display of *Aurora borealis* was the cause of a peculiar disturbance of the magnetic needle; and Dr. Dalton(178) was the first to show that the luminous rays of the Aurora are always parallel to the dipping-needle, and that the Auroral arches cross the magnetic meridian at right angles. Hansteen and Arago have attended with particular care to these influences of the Northern Lights, and the results of their observations are: —

That as the crown of the Aurora quits the usual place, the dipping-needle moves several degrees forward:

That the part of the sky where all the beams of the Aurora unite, is that to which a magnetic needle directs itself, when suspended by its centre of gravity:

That the concentric circles, which show themselves previously to the luminous beams, rest upon two points of the horizon equally distant from the magnetic meridian; and that the most elevated points of each arch are exactly in this meridian.(179)

It does not appear that every Aurora disturbs the magnetic needle; as Captains Foster and Back both describe very splendid displays of the phenomena, which did not appear to produce any tremor or deviation upon their instruments.(180)

Some sudden and violent movements have been from time to

time observed to take place in suspended magnets; and since the establishment of magnetic observatories in almost every part of the globe, a very remarkable coincidence in the time of these agitations has been detected. They are frequently connected with the appearance of Aurora borealis: but this is not constantly the case. These disturbances have been called *magnetic storms;* and over the Asiatic and European continent, the Islands of the Atlantic and the western hemisphere, they have been proved to be simultaneous.

From observations made at Petersburg by Kupffer, and deductions drawn from the observations obtained by the Magnetic Association, it appears probable that these *storms* arise from a sudden displacement in the magnetic lines of the earth's surface; but the cause to which this may be due is still to be sought for.

In the brief and hasty sketch which has been given of the phenomena of terrestrial magnetism, enough has been stated to show the vast importance of this very remarkable power in the great operations of nature. We are gradually reducing the immense mass of recorded observations, and arriving at certain laws which are found to prevail. Still, the origin of the force, whether it is strictly electrical, whether it is the circulation of magnetic fluid, or whether it is merely a peculiar excitation of some property of matter, are questions which are open for investigation.

In the beautiful Aurora borealis, with its trembling diffusive lights, and its many-coloured rays, we have what may be regarded as a natural exhibition of magnetism, and we appear to have within our grasp the explanation we desire. But we know not the secret of even these extraordinary meteorological displays. If we pass an electric spark from a machine through a long cylinder, exhausted of air as far as possible, we have a mimic representation of the Northern Lights — the same attenuation of brightness, almost dwindling into phosphorescence; and by the slightest change of temperature, we may produce that play of colours, which is sometimes so remarkably manifested in Aurora. Dr. Dalton considered Aurora borealis *as a magnetic phenomenon, and that its beams are governed by the earth's magnetism.* We know that the arc of light produced between the poles of a powerful galvanic battery is readily deflected by a good magnet; and we have lately learned that every vapour obeys the magnetic force.([181]) It is, therefore, yet a question for our consideration, does the earth's magnetism produce the peculiar phenomena of Aurora by acting upon electricity in a state of glow? or have we evidence in this display, of the circulation of the magnetic fluid around our globe, manifesting itself by its action on the ferruginous and other metallic matter, which Fusinieri has proved to exist in the upper regions of our atmosphere?([182])

The alteration in the properties of heat, when it passes from the radiant state into combination with matter, exhibits to us something like what we may suppose occurs in the conversion of magnetism into electricity or the contrary. We have a subtile agent, which evidently is forever busy in producing the necessary conditions of change in this our earth: an element to which is due the development of many of the most active powers of nature; performing its part by blending with those principles which we have already examined; associating itself with every form of matter; and giving, as we shall presently see, in all probability, the first impulses to combination, and regulating the forms of aggregating particles.

As electricity has the power of altering the physical conditions of the more adherent states of matter, thus giving rise to variations of form and modes of combination, so gross matter appears to alter the character of this agency, and thus dispose it to the several modifications under which we have already detected its presence. We have mechanical electricity and chemical electricity, each performing its great work in nature; yet both manifesting conditions so dissimilar, that tedious research was necessary before they could be declared identical. Magnetic electricity is a third form; all its characteristics are unlike the others, and the office it appears to perform in the laboratory of creation, is of a different order from that of the other states of electrical force. In the first two we have decomposing and recombining powers constantly manifested, in fact, their influences are always of a chemical character; but in the last it appears we have only a directive power. It was thought that evidence had been detected of a chemical influence in magnetism; it did appear that sometimes a retarding force was exerted, and often an accelerating one. This has been again denied, and we have arrayed in opposition to each other some of the first names among European experimentalists. The question is not yet to be regarded as settled; but, from long and tedious investigation, during which every old experiment has been repeated, and numerous new ones tried, we incline to the conclusion that chemical action is not directly affected by magnetic power. It is highly probable that magnetism may, by altering the structural arrangement of the surface, vary the rate of chemical action, but this requires confirmation.([183])

There is no substance to be found in nature existing independently of magnetic power. But it influences bodies in different ways: one set acting with relation to magnetism, like iron, and arranging themselves along the line of magnetic force, — these are called magnetic bodies; another set, of which bismuth may be taken as the representative, always placing themselves at right angles to this line, — these are called dia-magnetic bodies.([184]) This is strikingly shown by means of powerful electro-magnets, but the magnetism of the earth is sufficient, under proper care, to exhibit the phenomena.

Every substance in nature is in one or other of these conditions. The rocks, forming the crust of the earth, and the minerals which are discovered in them; the surface soil, which is by nature prepared as the fitting habitation of the vegetable world, and every tree, shrub, and herb which finds root therein, with their carbonaceous matter, in all its states of wood, leaf, flower, and fruit; the animal kingdom, from the lowest monad through the entire series up to man, — have, all of them, distinct magnetic or dia-magnetic relations.

"It is a curious sight," says Dr. Faraday, "to see a piece of wood or of beef, or an apple, or a bottle of water repelled by a magnet, or, taking the leaf of a tree, and hanging it up between the poles, to observe it take an equatorial position. Whether any similar effects occur in nature among the myriads of forms which, upon all parts of its surface, are surrounded by air, and are subject to the action of lines of magnetic force, is a question which can only be answered by future observation.(185)

At present, the bodies which are known to exhibit decided ferro-magnetic properties, are the following, which stand arranged in the order of their intensity: —

Iron, Nickel, Cobalt, Manganese,
Chromium, Cerium, Titanium,
Palladium, Platinum, Osmium.

It is interesting to know that there are evidences that two bodies which, when separate, are not magnetic, as iron is, become so when combined. Copper and zinc are both of the dia-magnetic class, but many kinds of brass are discovered to be magnetic.

The salts of the above metals are, to a greater or less extent, ferro-magnetic, but they may be rendered neutral by water, which is a dia-magnetic body, being repelled by the magnet. It will be unnecessary, here, to enumerate the class of bodies which are dia-magnetic; indeed, all not included in the preceding list may be considered as belonging to that class, with the exception of gases and vapours, which appear to exist, relatively to each other, sometimes in the one, and sometimes in the other condition.([186])

To endeavour to reduce our knowledge of these facts to some practical explanation, we must bear in mind that particular spaces around the north and south geographical poles of the earth are regarded as circles to which all the magnetic lines of force converge. Under circumstances which should prevent any interference with what is called ferro-magnetic action, all bodies coming under that class would arrange themselves, according to the laws which would regulate the disposition of an infinite number of magnets, free to move within the sphere of each other's influence. The north and south pole of one magnetic body would attach itself to the south and north pole of another, until we had a line of magnets of any extent; the two ends being in opposite states, like the magnetic points of convergence of the earth.

Every body, not ferro-magnetic, places itself across such a line of magnetic force as we have conceived; and if the earth was made up of separate layers of ferro-magnetic and dia-magnetic bodies, the result would be the formation of bands at right angles to each other. This is not the case, by reason of the intermingling of the two classes of substances. Out of the known chemical elements we find only about ten which are actively ferro-magnetic; the others combining with these give rise to either a weaker state, a neutral condition, or the balance of action is turned to the dia-magnetic side. Sulphate of iron, for instance, is a magnetic salt; but in solution, water being dia-magnetic, it loses its property. The yellow prussiate of potash is a dia-magnetic body; but the red prussiate, which contains an atom less of potassium, is magnetic.([187])

These two conditions of matter stand, therefore, in opposition;

and as every particle of any substance found in this earth is endued with the property of disposing itself according to one or the other of these powers, we appear to be approaching to a knowledge of the causes of molecular arrangements.

We still search in the dark, and see but dimly the evidences; yet it becomes almost a certainty to us, that this stone of granite, with its curious arrangement of felspar, mica, and quartz, presents its peculiar condition in virtue of some such law as that of dia-magnetism. The crystal, too, of quartz, which we break out of the mass, and which presents to us a beautifully regular figure, is, beyond a doubt, so formed, because the atoms of silica are each one impelled in obedience to one of these two forms of magnetism to set themselves in a certain order to each other, which cannot be altered by human force without destruction.

All the laws which regulate the forms of crystals and amorphous bodies are, to the greatest degree, simple. In nature, the end is ever attained by the easiest means; and the complexity of operation, which appears sometimes to the observer, is only so because he cannot see the spring by which the machine is moved.

The gaseous envelope, our atmosphere, is in a neutral state. Oxygen is strikingly magnetic in relation to hydrogen gas, whilst nitrogen is as singularly the contrary; and the same contrasts present themselves when these gases are examined in their relation to common air. Thus, oxygen being magnetic, and nitrogen the contrary, we have an equilibrium established, and the result is a compound, neutral in its relations to all matter. All gases and vapours are found to be dia-magnetic, but in different degrees.(18) This is shown by passing a stream of the gas, rendered visible by a little smoke, within the influence of a powerful magnet. These bodies are, however, found relatively to each other,—or even to themselves, under different thermic conditions, — to change their states, and pass from the magnetic to the dia-magnetic class.

Heat has, however, a very remarkable influence in altering these relations; and atmospheric air at one temperature is magnetic to the same fluid at another: thus, by thermic variations, attraction or repulsion is alternately maintained. By this it must be understood that a stream of air, at a temperature elevated but a few degrees above that of an atmosphere of the same kind into which it is passing, is deflected in one way by a magnet; whereas, if the stream is colder than the bulk through which it flows, it is bent in another way by the same force. In this respect, magnetism and dia-magnetism show equally the influence of another physical force, heat; and we may safely refer many meteorological phenomena to similar alterations of condition in the atmosphere, relative to the magnetic relations of the aërial currents.

That magnetism has a directive power is satisfactorily shown by the formation of crystals in the neighbourhood of the poles of powerful magnets. The common iron salt, the proto-sulphate, ordinarily crystallizes, so that the crystals unite by their faces; but, when crystallizing under magnetic influence, they have a tendency to arrange themselves with regard to each other, so that the acute angle of one crystal unites with one of the faces of another crystal, near to, but never actually at, its obtuse angle. In addition to this, if a magnet of sufficient power is employed, the crystals arrange themselves in magnetic curves from one pole to the other, a larger crop of crystals being always formed at the north than at the south pole. Here we have evidence of an actual turning round of the crystal, in obedience to the directive force of the magnet; and we have the curious circumstance of a difference in some way, which is not clearly explained, between the two opposite poles. If, instead of an iron, or a ferro-magnetic salt, we employ one which belongs to the other, or dia-magnetic class, we have a curious difference in the result. If into a glass dish, fixed on the poles of a strong electro-magnet, we pour a quantity of a solution of nitrate of silver, and place in the fluid, over the poles of the magnet, two globules of mercury, by which that arborescent crystallization, called the *Arbor Dianæ*, is produced, we have the long needle-shaped crystals of silver, arranging themselves in curves which would cut the ordinary magnetic lines at right angles.([189])

In the first example given, we have an exhibition of magnetic force, while in the last we have a striking display of the dia-magnetic power.

The large majority of natural formations appear to group themselves under the class of dia-magnetics. These bodies are thought to possess poles of mutual repulsion among themselves, and which are equally repelled by the magnetic points of convergence. Confining our ideas to single particles in one condition or the other, we shall, to a certain extent, comprehend the manifold results which must arise from the exercise of these two modes of force. At present, our knowledge of the laws of magnetism is too limited to allow of our making any general deductions relative to the disposition of the molecules of matter; and the amount of observation which has been given to the great natural arrangements, is too confined to enable us to infer more than that it is probable many of the structural conditions of our planet are due to some polar action.

Mountain ranges observe a singular uniformity of direction, and the cleavage planes of rock are evidently due to some all-pervading power. Mineral bodies are not distributed in all rocks indiscriminately. The primary formations hold one class of metalliferous ores, and the more recent ones another. This is not to be regarded as in any way connected with their respective ages, but with some peculiar condition of the stone itself. The granite and slate rocks, at their junctions, present the required conditions for the deposit of copper ore, while we find the limestones have the characteristic physical state for accumulating lead ore. Again, on examining any mineral vein, it will be at once apparent that every particle of ore, and every crystal of quartz or limestone, is disposed in a direction which indicates the exercise of some powerful directive agency.(100)

It appears, from all the results hitherto obtained, that the magnetic and dia-magnetic condition of bodies is equally due to some peculiar property of matter in relation to the other forms of electricity. We have not yet arrived at the connecting link, but it does not appear to be far distant.

We have already referred to the statement made by talented experimentalists, that magnetism has a powerful influence in either retarding or accelerating chemical combination. Beyond a doubt chemical action weakens the power of a magnet; but the disturbance which it occasions in soft iron, on the contrary, appears to tend to its receiving magnetism more readily, and retaining it more permanently. Further investigations are, however, required, before we can decide satisfactorily either of these problems, both of which bear very strongly upon the subject we have just been considering.

We have seen that heat and electricity act strangely on magnetic force, and that this statical power reacts upon them; and thus the question naturally arises, Do light and magnetism in any way act upon each other?

Morichini and Carpi on the continent, and Mrs. Somerville in England, have stated that small bars of steel can be rendered magnetic by exposing them to the influence of the violet rays of light. These results have been denied by others, but again repeated and apparently confirmed. In all probability, the rays to which the needles were exposed, being those in which the maximum actinic power is found, produced an actual chemical change; and then, if the position were favourable, it is quite evident that magnetism would be imparted. Indeed, we have found this to be the case when the needles, exposed to solar radiations, were placed in the direction of the dip. The supposed magnetization of light by Faraday has already been mentioned. If the influence in one case is determined, it will render the other more probable.([191])

"In seeking for a cause," writes Sir David Brewster, "which is capable of inducing magnetism on the ferruginous matter of our globe, whether we place it within the earth, or in its atmosphere, we are limited to the SUN, to which all the magnetic phenomena have a distinct reference; but, whether it acts by its heat, or by its light, or by specific rays, or influences of a magnetic nature, must be left to future inquiry.([192])

We have learnt that magnetism is not limited to ferruginous matter; we know that the ancient doctrine of the universality of the property is true. Kircher, in his strange work on Magnetism published in the early part of the seventeenth century([193]) — a curious exemplification of the most unwearying industry and careful experiment, combined with the influences of the credulity and superstitions of his age — attributes to this power nearly all the cosmical phenomena with which, in his time, men were acquainted. He curiously anticipates the use of the supposed virtue of magnetic traction in the curative art; and as the titles of his concluding chapters sufficiently show, he was a firm believer in animal magnetism.([194]) But it is not with any reference to these that we refer to the work of *Athanasii Kircheri, Societatis Jesu, Magnes, sive de Magnetivâ Arte,* but to show that, two hundred years since, man was near a great truth; but the time of its development being not yet come, it was allowed to sleep for more than two centuries, and the shadow of night had covered it. In speaking of the vegetable world, and the remarkable processes by which the leaf, the flower, and the fruit are produced, this old sage brings forward the fact of the dia-magnetic character of the plant, which has been, within the last two years, re-discovered; and he refers the motions of the Sun-flower, the closing of the Convolvulus, and the directions of the spiral, formed by twining plants, to this particular influence.

This does not appear as a mere speculation, a random guess, but is the result of deductions from experiment and observation. Kircher doubtless leaped over a wide space to come to his conclusion; but the result is valuable in a twofold sense. In the first it shows us that, by neglecting a fact which is suggestive, we probably lose a truth of great general application; and secondly, it proves to us, that, by stepping beyond the point to which inductive logic leads, and venturing on the wide sea of hypothesis, we are liable to sacrifice the true to the false, and thus to hinder the progress of human knowledge.

Magnetism, in one or other of its forms, is now proved to be universal, and to its power we are disposed to refer the structural conditions of all material bodies, both organic and inorganic. This view has scarcely yet been recognized by philosophers; but as we find a certain law of polarity prevailing through every atom of created matter, in whatever state it may be presented to our senses, it is evident that every particle must have a polar and directing influence upon the mass, and every coherent mass becomes thus only a larger and more powerful representative of the magnetic unit. Thus we see the speculation of Hansteen, that the sun is, to us, a magnetic centre, and that it is equally influenced by the remoter suns of the universe,([105]) is supported by legitimate deductions from experiment.

The great difficulty is not, however, got rid of by this speculation; the cause by which the earth's magnetism is induced is only removed further off.

The idea of a magnetic fluid is scarcely tenable; and the ferruginous nature of the Aurora borealis receives no proof from any investigation; indeed, we have procured evidence to show that iron is not at all necessary for the production of magnetic phenomena. The leaf of a tree, a flower, fruit, a piece of animal muscle, glass, paper, and a variety of similar substances, have the power of repelling the bar of iron which we call a magnet, and of placing it at right angles to the direction of the force exerted by them. This is a point which must be constantly borne in mind, when we now consider the mysteries of magnetic phenomena.

Any two masses of matter act upon each other according to this law, and although by the power of cohesion the force may be brought to an equilibrium, or to its zero point, it is never lost, and may be readily and rapidly manifested by any of the means employed for electrical excitation.

Reasoning by analogy, the question fairly suggests itself: If two systems of inorganic atomic constitution are thus invested with a power of influencing each other through a distance, why may not two more highly developed organic systems equally, or to a greater extent, produce an influence in like manner? Upon such reasoning as this is founded the phenomenon known as Animal Magnetism. There is no denying the fact that one mass of blood, muscle, nerves, and bone, must, magnetically, influence another similar mass. This is, however, something totally different from that abnormal condition, which is produced through some peculiar and, as yet, unexplained physiological influences.

With the mysterious operations of vital action, the forces which we have been considering have nothing whatever in common. The powers which are employed in the arrangements of matter are, notwithstanding their subtile character, of far too gross a nature to influence the psychological mysteries which present themselves to the observant mind. It cannot be denied that, by placing a person of even moderate nervous sensibility in the constrained position, and under the unnatural influence of the mind, as acquired by the disciples of Mesmer, a torpor affecting only certain senses is produced. The recognized and undoubted phenomena are in the highest degree curious — but in these the marvels of charlatanry and ignorance are not included; — and the explanation must be sought for by the physiologist among those hidden principles upon which depends all human sensation. ([196])

Man, like a magician, stands upon a promontory, and, surveying the great ocean of the physical forces which involve the material creation, and produce that infinite variety of phenomena which is unceasingly exhibited around him, he extends the wand of intelligence, and bids the "spirits of the vasty deep" obey his evocation.

The phenomena recur — the great processes of creation go on — the external manifestations of omnipotent power proceed — effects are again and again produced; but the current of force passes undulating onwards; — and to the proud bidding of the evocator there is no reply but the echo of his own vain voice, which is lost at last in the vast immensity of the unknown which lies beyond him.

We see how powerfully the physical forces, in their various modes of action, stir and animate this planetary mass; and amongst these the influence of magnetism appears as a great directing agent, though its origin is unknown to us.

That power which, like a potent spirit, guides
The sea-wide wanderers over distant tides,
Inspiring confidence where'er they roam,
By indicating still the pathway home; —
Through nature, quickened by the solar beam,
Invests each atom with a force supreme,
Directs the caverned crystal in its birth,
And frames the mightiest mountains of the earth;
Each leaf and flower by its strong law restrains,
And man, the monarch, binds in iron chains.

CHAPTER XI.

CHEMICAL FORCES.

Nature's Chemistry — Changes produced by Chemical Combination — Atomic Constitution of Bodies — Laws of Combination — Combining Equivalents — Elective Affinity — Chemical Decomposition — Compound Character of Chemical Phenomena — Catalysis, or action of Presence — Transformation of Organic Bodies — Organic Chemistry — Constancy of Combining Proportions — The Law of Volumes, the Law of Substitutions, Isomeric States, &c.

All things on the earth are the result of chemical combination. The operations by which the commingling of molecules and the interchange of atoms take place, we can imitate in our laboratories; but in nature they proceed by slow degrees, and, in general, in our hands they are distinguished by suddenness of action. In nature, chemical power is distributed over a long period of time, and the process of change is scarcely to be observed. By art we concentrate chemical force, and expend it in producing a change which occupies but a few hours at most. Many of the more striking phenomena of nature are still mysterious to us, and principally because we do not, or cannot, take the element time into calculation. The geologist is compelled to do this to explain the progress of the formation of the crust of the earth, but the chemist rarely regards the effects of time in any of his operations. The chemical change which within the fissure of the rock is slowly and silently at work, displacing one element or molecule, and replacing it by another, is in all probability the operation of a truly geological period. Many, however, of the changes which are constantly going on around us, are of a much more rapid character, and in these nature is no slower in manipulating than the chemist.

Had it been that the elements which are now found in combination could exist in a free state, the most disastrous consequences would necessarily ensue. There must have been a period when many of the combinations known to us were not yet created. Their elements either existed in other forms, or were uncombined. Our rocks are compounds of oxygen with certain peculiar metals which unite with oxygen so rapidly that incandescence is produced by their combination. Let us suppose that any of these metals existed in purity, and that they were suddenly brought into contact with water, the atmospheric air, or any body containing oxygen, the result would be a convulsion of the most fearful kind; the entire mass of metal would glow with intensity of heat, and the impetuosity of the action would only be subdued when the whole of the metal had become oxidized. Volcanic action has been referred to some such cause as this, but there is not sufficient evidence to support the hypothesis; indeed, the opinion of most philosophers is against it.([197]) Such a condition may possibly have existed at one time, but it is only adduced here as an example of the violent nature of some chemical changes. Potassium thrown on water bursts into flame, and sodium does so under certain conditions. If these, or the metals proper in a state of fine division, are brought into an atmosphere of chlorine, the intensity of chemical action is so great that they become incandescent, many of them glowing with extreme brilliancy. If hydrogen gas is mixed with this element (chlorine) they unite, under the influence of light, with explosive violence, giving rise to a compound, muriatic acid, which combines with water in an almost equally energetic manner. Nitrogen, as it exists in the atmosphere, mixed with oxygen, appears nearly inert; with hydrogen it forms the pungent compound ammonia;

with carbon, the poisonous one cyanogen, the base of Prussic acid; with chlorine it gives rise to a fluid, oily in its appearance, but which, when merely touched by an unctuous body, explodes more violently than any other known compound, shivering whatever vessel it may be contained in, to atoms; with iodine it is only slightly less violent; and in certain combinations with silver, mercury, gold, or platinum, it produces fulminating compounds of the most dangerous character.(196) Here we have elements harmless when uncombined, exhibiting the most destructive effects if their combinations are at all disturbed; and in the other case we have inert masses produced from active and injurious agents.

We regard a certain number of substances as *elementary;* that is to say, not being able, in the present state of our knowledge, to reduce them to any more simple condition, they are regarded as the elements, which by combination produce the variety of substances found in the three kingdoms of nature.

We have already spoken of the atomic constitution of bodies. It remains now to explain the simplicity and beauty, which mark every variety of combination under chemical force. As a prominent and striking example, water is a compound of two gaseous bodies, oxygen and hydrogen: —

If we decompose water by means of galvanic electricity, we shall find the volume of hydrogen gas to be to that of the oxygen as two to one; and if equal volumes of those gases are weighed, the oxygen will be found to be sixteen times heavier than the hydrogen. Hence we infer that water, which is

Oxygen	100
Hydrogen	12.5

is composed of one atom of hydrogen gas weighing 1, and an atom of oxygen weighing 8, relatively to the hydrogen. It is found in the same way that the theoretical weight of the atom of carbon is 6, and that of nitrogen 14; whilst the atom of iron is 28, that of silver 108, of gold 199, and that of platinum and iridium each 98.([199]) Now, as these are the relative weights of the ultimate indivisible atom, it follows that all combinations must be either atom to atom, or one to two, three, or four; but that in no case combination should take place in any other than a multiple proportion of the equivalent or atomic number. This is found to be the case. Oxygen, for instance, combines, as one, two, or three atoms: its combination presenting some multiple of its equivalent number 8, as 16, or 24; and in like manner the combining quantity of carbon is 6, or some multiple of that number. Where this law is not found strictly to agree with analytical results, of which some examples are afforded by the sesquioxides, it may be attributed, without doubt, to some error of analysis, or in the method of calculation.

Nothing can be more perfect than the manner in which nature regulates the order of combination. We have no uncertain arrangement; but, however great the number of the atoms of one element may be, over those of another, those only combine which are required, according to this great natural law, to form the compound, all the others still remaining free and uncombined. These results certainly appear to prove that the elementary particles of matter are not of the same specific gravities. Do they not also indicate that any alteration in the specific gravity of the atom would give rise to a new series of compounds, thus apparently producing a new element? Surely there is nothing irrational in the idea that the influences of heat or electricity, or of other powers of which as yet we know nothing, may be sufficient to effect such changes in the atomic constituents of this earth.

The combination of elementary atoms takes place under the influence of an unknown force, which we are compelled to express by a figurative term, *affinity*. In some cases it would appear that the disposition of two bodies to unite, is determined by the electrical condition; but a closer examination of the question than it is possible to enter into in this place, clearly shows that some physical state, not electrical, influences combining power.

Chemical affinity or attraction is the peculiar disposition which one body has to unite with another. To give some instances in illustration, water and spirit combine most readily: they have a strong affinity for each other. Water and oil repel each other: they will not enter into combination. If carbonate of potash is added to the spirit and water in sufficient quantity, the water is entirely separated, and the pure spirit will float over the hydrated potash. If potash is added to the oil and water, it combines with the oil, and, forming soap, they all unite together; but, if we now add a little acid to the mixture, the potash will quit the oil to combine with the acid, and the oil will be repelled as before, and float on the liquid. This has been called single elective affinity. These elections were regarded as constant, and chemists drew up tables for the purpose of showing the order in which these decompositions occur.([200]) Thus, ammonia, it was shown, would separate sulphuric acid from magnesia, lime remove it from ammonia, potash or soda from lime, and barytes from potash or soda. It was thought the inverse of this order would not take place, but recent researches have shown that the results are modified by quantity and some other conditions.

It often happens that we have a compound action of this kind in which double election is indicated. Sulphate of lime and carbonate of ammonia in solution are brought together, and there results a carbonate of lime and a sulphate of ammonia. Now, in such cases nothing more than single elective attraction most probably occurs, and the carbonic acid is seized by the lime, only after it has been set free from the ammonia, by the great affinity of that earth for carbonic acid: and then, by the force of cohesion acting with the combining powers, the insoluble salt is precipitated.(201) There is a curious fact in connection with this decomposition. If carbonate of lime and sulphate of ammonia are mixed together dry, and exposed, in a closed vessel, to a red heat, sulphate of lime and carbonate of ammonia are formed. These opposite effects are not very easily explained. The action of heat is to set free the carbonic acid; and it can only be by supposing that considerable differences of temperature reverse the laws of affinity, that we can at all understand this phenomenon. That different effects result at high temperatures from those which prevail at low ones, recent experiments prove to us, particularly those of Boutigny already quoted, when considering decomposition by calorific action.

Under the term chemical affinity, which we regard as a power acting at insensible distances, and producing a change in bodies, we are content to allow ourselves to believe that we have explained the great operations of nature. We find that the vegetable and animal kingdoms are composed of carbon, hydrogen, oxygen, and nitrogen. The granite mountains of the earth, and its limestone hills, and all its other geological formations, are found to be metals and oxygen, and carbon and sulphur, disposed to settle in harmonious union in their proper places by chemical affinity. But what really is the power which combines atom to atom, and unites molecule to molecule? Can we refer the process to heat? The influence of caloric, although by changing the form of bodies it sometimes assists combination, is to be regarded rather as in antagonism to the power of cohesion. Can it be thought that electricity is active in producing the result? During every change of state, those phenomena, which we term electrical, are manifested; but we thereby only prove the general diffusion of the electric principle, and by no means show that electricity is the cause of the chemical change. Can light determine these changes? It is evident, although light may be a disturbing power, that it cannot be the effective one; for many of these decompositions and recompositions are constantly going on within the dark and silent depths of the earth, to which a sunbeam cannot reach. That the excitation on the surface of the earth, produced by solar influence, may modify those changes, is probable. It is, however, certain that we must regard all manifestations of chemical force as dependent upon some secret principles common to all matter, diffused throughout the universe, but modified by the influences of the known imponderable elements, and by the mechanical force of aggregative attraction.

Bodies undergo remarkable changes of form, and present very different characters, by re-actions, which are of several kinds. We suppose that a permanent corpuscular arrangement is maintained so long as the equilibrium of the molecular forces is undisturbed. Water, for instance, remains unchanged so long as the balance of affinity is kept up between the oxygen and hydrogen of which it is composed, or so long as the oscillations of force between these combining elements are equal; but disturb this force, or set up a new vibratory action, as by passing an electric current through the water, or by presenting another body, which has the power of re-acting upon one of these corpuscular systems, and the water is decomposed, the hydrogen and oxygen gases being set free, or one alone is liberated, and the other combined with the molecules of the agent employed, and a new compound produced. This is chemistry, by which science we discover the laws which regulate all combinations of matter.

Having reason to conclude that atom combines with atom, according to a system most harmoniously arranged, there can be no difficulty in conceiving that molecule unites with molecule, in a manner regulated by some equally well-marked law. It was, indeed, a discovery by Wenzel, of Fribourg, that, in salts which decompose each other, the acid which saturates one base will also saturate the other base; and the subsequent observations of Richter, of Berlin, who attached proportional numbers to the acids and bases, and who remarked that the neutrality of metallic salts does not change during the precipitation of metals by each other, which led the way to the atomic theory of Dr. Dalton, to whom entirely belongs the observation, that the equivalent of a compound body is the sum of the equivalents of its constituents, and the discovery of combination in multiple proportions.

The elements of a molecule can take a new arrangement amongst themselves, without any alteration in the number of the atoms or of their weight, and thus give rise to a body of a different form and colour, although possessing the same chemical constitution. This is the case with many of the organic compounds of carbon and hydrogen.

The elements of a compound may be dissociated, and thus the dissimilar substances of which it is composed, set free. A piece of chalk exposed to heat is, by the disturbance of its molecular arrangement, changed in its nature; a gaseous body, carbonic acid, is liberated, and quicklime (oxide of calcium) is left behind. If this carbonic acid is passed through red-hot metal tubes, or brought in contact with heated potassium, it is resolved into oxygen and charcoal — the oxygen combining with the metal employed. The oxide of calcium (lime), if subjected to the action of a powerful galvanic current, is converted into oxygen and a metal, calcium. Thus we learn that chalk is a body consisting of two compound molecules, — carbonic acid, which is formed by the combination of an atom of carbon with two atoms of oxygen, — and lime, which results from the union of an atom of calcium with one of oxygen.

The condition requisite to the production of chemical action between bodies is that they should be dissimilar. Two elementary atoms are placed within the spheres of each other's influences, and a compound molecule results. Oxygen and hydrogen form water, oxygen and carbon give rise to carbonic acid; nitrogen and hydrogen unite to form ammonia; and cholorine and hydrogen to produce hydrochloric acid. In all these cases an external force is required to bring the atoms within the range of mutual affinity: flame, the electrical spark, actinism, or the interposition of a third body, is necessary in each case. There are other examples in which no such influence is required. Potassium and oxygen instantly unite; chlorine, iodine, and bromine immediately, and with much violence, combine with the metals to form chlorides, iodides, or bromides.

With compound molecules the action is in many cases equally active, and combination is readily effected, as in the cases of the acids and the oxides of some metals, which are all instances of the most common chemical attraction.

An elementary or simple molecule and molecules of a compound and different constitution are brought together, and a new compound results from an interchange of their atoms, whilst an element is liberated. These are essentially illustrations of analytical chemistry. Sulphuretted hydrogen is mixed with chlorine; the chlorine combines with the hydrogen, and sulphur is set free. Potassium is put into water, and it combines with the oxygen of the water, whilst the hydrogen is liberated.

Two compound molecules being brought together may decompose each other, and form two new compounds by an interchange of their elements.

One element may be substituted for another under certain circumstances. Gold may be replaced by mercury; copper will take the place of silver; and iron will occasion the separation of copper from its solutions, the iron itself being dissolved to supply its place; chlorine will substitute hydrogen in the carburetted hydrogen gases; and many other examples might be adduced.

Chemical phenomena very frequently become of a complex character; and one, two, or three of these cases may be occurring at the same time in the decomposition of one compound by another. Such are the general features of chemical science. Many peculiarities and remarkable phenomena connected with chemical investigations will be named, as the examination of the elementary composition of matter is proceeded with; but, although the philosophy of chemical action is of the highest interest, it must not be allowed to detain us with its details, which are, indeed, more in accordance with a treatise on the science than one which professes to do no more than sketch out those prevailing and striking features which, whilst they elucidate the great truths of nature, are capable of being employed as suggestive examples of the tendency of scientific investigation to enlarge the boundaries of thought, and give a greater elevation to the mind, leading us from the merely mechanical process of analysis up to the great synthetical operations, by which all that is found upon the earth for its ornament or our necessities, is created.

Among the most remarkable phenomena within the range of physical chemistry are those of *Catalysis*, or, as it has also been called, the "*Action of Presence.*"([202]) There are a certain number of bodies known to possess the power of resolving compounds into new forms, without undergoing any change themselves. Kirchoff discovered that the presence of an acid, at a certain temperature, converted starch into sugar and gum, no combination with the acid taking place. Thenard found that manganese, platinum, gold, and silver, and, indeed, almost any solid organic body, had the power of decomposing the binoxide of hydrogen, by their presence merely, no action being detected on these bodies. Edmund Davy found that powdered platinum, moistened with alcohol, became red-hot, fired the spirit, and converted it into vinegar, without undergoing, itself, any chemical change. Dœbereiner next discovered that spongy platinum fired a current of hydrogen gas directed upon it, which, by combining with the oxygen of the air, formed water. Dulong and Thenard traced the same property, differing only in degree, through iridium, osmium, palladium, gold, silver, and even glass. Further investigation has extended the number of instances; and it has even been found that a polished plate of platinum has the power of condensing hydrogen and oxygen so forcibly upon its surface, that they are drawn into combination and form water, with a development of heat sufficient to ignite the metal.

This power, whatever it may be, is common in both organic and inorganic nature, and on its important purposes Berzelius has the following remarks: —

"This power gives rise to numerous applications in organic nature; thus, it is only around the eyes of the potato that diastase exists: it is by means of catalytic power that diastase, and that starch, which is insoluble, is converted into sugar and into gum, which, being soluble, form the sap that rises in the germs of the potato. This evident example of the action of catalytic power in an organic secretion, is not, probably, the only one in the animal and vegetable kingdom, and it may hereafter be discovered that it is by an action analogous to that of catalytic power, that the secretion of such different bodies is produced, all which are supplied by the same matter, the sap in plants, and the blood in animals."([263])

It is, without doubt, to this peculiar agency that we must attribute the abnormal actions produced in the blood of living animals by the addition of any gaseous miasma or putrid matter, of which we have, in all probability, a fearful example in the recent progress of Asiatic Cholera; therefore the study of its phenomena becomes an important part of public hygiène.

Physical research has proved to us that all bodies have peculiar powers, by which they condense with varying degrees of force gases and vapours upon their surfaces; every body in nature, may, indeed, be regarded as forming its own peculiar atmosphere. To this power, in all probability, does catalysis belong. Different views have, however, prevailed on this subject, and recently Dr. Lyon Playfair has published an elaborate Memoir,(204) in which he argues on the probability that the catalytic force is merely a modified form of chemical affinity, exerted under peculiar conditions.

Whatever may be the power producing chemical change, it acts in conformity with some fixed laws, and, in all its transmutations, an obedience to a most harmonious system is apparent.

It is curious to observe the remarkable character of many of these natural transmutations of matter, but we must content ourselves with a few examples only. For instance: —

Sugar, oxalic acid, and citric acid are very unlike each other, yet they are composed of the same elements; the first is used as a general condiment, the second a destructive poison, and the third a grateful and healthful acid: sugar is readily converted into oxalic acid, and, in the process of ripening fruits, Nature herself converts citric acid into sugar. Again, starch, sugar, and gum would scarcely be regarded as alike, yet their only difference is in the mode in which carbon, hydrogen, and oxygen combine. They are composed of the same principles, in the following proportions: —

	Carbon.	Hydrogen.	Oxygen.
Starch . . .	12 ..	10 ..	10
Sugar . . .	12 ..	11 ..	11
Gum . . .	12 ..	11 ..	11

These *isomeric* groups certainly indicate some law of affinity which science has not yet discovered. Similar and even more remarkable instances might be adduced of the same elements producing compounds very unlike each other; but the above have been selected from their well-known characters. Indeed, we may state with truth that all the varieties of the vegetable world — their woody fibre — their acid or alkaline juices — the various exudations of plants — their flowers, fruit, and seeds, and the numerous products which, by art, they are made to yield for the uses of man, are, all of them, compounds of these three elements, differing only in the proportions in which they are combined with nitrogen, or in some peculiar change of state in one or other of the elementary principles. By the impulse given to organic chemistry by Liebig, our knowledge of the almost infinite variety of substances, in physical character exceedingly dissimilar, which result from the combination of oxygen, hydrogen, and carbon, in varying proportions, has been largely increased. And the science is now in that state which almost causes a regret that any new organic compounds should be discovered, until some industrious mind has undertaken the task of reducing to a good general classification, the immense mass of valuable matter which has been accumulated, but which, for all practical purposes, remains nearly useless and unintelligible.

These combinations, almost infinitely varied as they are, and so readily produced and multiplied as to be nearly at the will of the organic analyst, are not, any of them, accidental: they are the result of certain laws, and atom has united with atom in direct obedience to principles which have been through all time in active operation. They are unknown; the researches of science have not yet developed them, and the philosopher has not yet made his deductions. They are to be referred to some secret fixed principles of action, to a force which has impressed upon every atom of the universe its distinguishing character. Chemistry makes us familiar with a system of order. The researches of analysts have proved that every body has a particular law of combination, to which it is bound by a mathematical precision; but it is not proportional combination alone we have to consider. If *allotrophy* is evidenced in the mineral world, it is certainly far more strikingly manifested in the vegetable and animal kingdoms.

There are some cases in which bodies appear to combine without any limitation, as spirit of wine and water, sulphuric acid and water; but these must be considered as conditions of mixture rather than of chemical combination.

The composition of bodies is fixed and invariable, and a compound substance, so long as it retains its characteristic properties, must consist of the same elements united in the same proportions. Thus, sulphuric acid is invariably composed of 16 parts of sulphur and 24 parts of oxygen. Chalk, whether formed by nature or by the chemist, yields 43.71 parts of carbonic acid, and 56.29 parts of lime. The rust which forms upon the surface of iron by the action of the atmosphere, is as invariable in its composition as if it had been formed by the most delicate adjustment of weight by the most accurate manipulator, being 28 parts of iron and 12 parts of oxygen. This law is the basis of all chemical inquiry, all analytical investigations depending upon the knowledge it affords us, that we can only produce certain undeviating compounds as the results of our decompositions. We are not in a position to offer any explanation which will account for these constant quantities in combination. The forces of cohesion and elasticity have been advanced in explanation, on the strength of the fact that the solubility of a salt in water is regulated by cohesion, and that of a gas by its elasticity. Although it may appear that some cases of chemical combination are due to these powers, — as, for instance, when the union of oxalic acid or sulphuric acid with lime produces an insoluble salt, — we cannot thus explain the constant proportions in which the metals, sulphur, oxygen, and similar bodies unite.

Another law teaches us, when compound bodies combine in more than one proportion, that every additional union represents a multiple of the combining proportion of the first; — with the difficulty which arises from the sub-multiple compounds we cannot deal; — further research may render their laws less obscure. We have seen that 8 parts of oxygen unite with 1 of hydrogen and 14 of nitrogen. It also unites with 110 of silver, 96 of platinum, 40 of potassium, 36 of chlorine, and 200 parts of mercury, giving rise to —

Water	9
Nitrous oxide	22
Oxide of silver	118
Oxide of platinum	104
Potash	48
Oxide of chlorine	44
Oxide of mercury	208

In these proportions, or in multiples of them, and in no others, will these bodies unite with the acids or other compounds. It will, of course, be understood that any other numbers may be adopted, provided they stand in the same relation to each other. ([205])

From the discovery of these harmonious arrangements was deduced the beautiful atomic theory to which allusion has been already made. Indeed, there does not appear to be any other way of explaining these phenomena, than by the hypothesis that the ultimate atoms of bodies have relatively the weights which we arbitrarily assign to them, as their combining quantities. These views are further confirmed by the fact, that gaseous bodies unite together by volume in very simple definite proportions: 100 measures of hydrogen and 200 measures of oxygen form water; 100 measures of oxygen and 100 measures of vapour of sulphur form sulphurous acid gas. Ammoniacal gas consists of 300 volumes of hydrogen and 100 volumes of nitrogen, condensed by combination into 200 volumes; consequently, we are enabled most readily to calculate the specific gravity of ammoniacal gas. The specific gravity of nitrogen is 0.9722, that of hydrogen 0.0694. Now, three volumes of hydrogen are equal to 0.2082, which added to 0.9722 is equal to 1.1804, which is exactly the specific gravity obtained by experiment.

There is no doubt, from the generality with which this law of volumes prevails, that it would be found to extend through all substances, provided they could be rendered gaseous; in other words, there is abundant proof to convince us that, throughout nature, the process of combination, in the most simple ratio of volumes, is in operation to produce all the forms of matter known to us.

It has been shown, by the admirable investigations on Atomic Volumes of Dr. Playfair and Mr. Joule, that salts, containing water of crystallization, dissolve in water without increasing its bulk more than is due to the liquefaction of the water which these salts contain, the anhydrous salts taking up no space in solution. This was first observed by Dr. Dalton, who, in 1840, remarked that sugar and certain hydrated salts, on solution in water, increased its volume by a quantity precisely equal to the volume of water they held in combination. From this we are naturally led to conclude, that the volume occupied by a salt in the solid state has a certain relation to the volume of the same salt when in solution, and has also a fixed relation to the volume occupied by any other salt. The law appears to be: the atomic volume of any salt whatever (anhydrous or hydrated) is a multiple of 11, or of a number near 11, or a multiple of 9.8 (the atomic volume of ice), or the sum of a multiple of 11 or 9.8. Marignac, who has also paid much attention to the subject, does not think these numbers absolutely correct, but approximately so.([206])

In addition to the laws already indicated, there appear to be some others of which, as yet, we have a less satisfactory knowledge, and, as a remarkable case, we may adduce the phenomena of *substitution*, or that power, which an elementary body, under certain conditions, possesses, of turning out one of the elements of a compound, and of taking its place.([207])

Under the influence of these laws, all the combinations which we discover in nature take place. The metals, and oxygen, and sulphur, and phosphorus unite. The elements of the organic type, entering into the closest relations, give rise to every form of vegetable life. The acids, the gums, the resins, and the sugar which plants produce; and those yet more complicated animal substances, bone, muscle, blood, and bile; albumen, casein, milk, with those compounds which, under the influence of vital power, resolve themselves into substances which are essential to the existence, health, and beauty of the animal fabric, are all dependent on these laws. But these metamorphoses must be further considered in our examination of the more striking cases of chemical action. The changes which result from organic combination are so remarkable, and withal they show how completely the whole of the material world is in subjection to chemical force, and every variety of form the result of mysterious combination, that some particular reference to these metamorphoses is demanded.

In nearly all cases of decided chemical action, every trace of the characters of the combining bodies disappear. We say decided chemical action, because, although sulphuric acid and water combine, and salts dissolve in water, we may always recognize their presence, and therefore these and similar cases cannot be regarded as strict examples of the phenomena under consideration.

Hydrogen and oxygen, in combining, lose their gaseous forms, and are condensed into a liquid — water. — Sulphuric acid is intensely sour and corrosive; potash is highly caustic; but united they form a salt which is neither: they appear to have destroyed the distinguishing characters of each other. Combined bodies frequently occupy less space than they did previously to combination, of which numerous particular instances might be adduced. Gases in many cases undergo a remarkable condensation when chemically combined. In slaking lime, the water becomes solid in the molecules of the hydrate formed, and the intense heat produced arises from the liberation of that caloric which had been employed to keep the water liquid. When a solid passes into the liquid state, cold is produced by the abstraction from surrounding objects of the heat required to effect fluidity. An alteration of temperature occurs whenever chemical change takes place, as we have already shown, with a few slight exceptions; and the disturbance caused by the exercise of the force of affinity frequently leads to the development of several physical powers.

Changes of colour very frequently arise; indeed, there does not appear to be any relation between the colour of a compound and that of its elements. Iodine is of a deep iron-gray colour; its vapour is violet; it forms beautifully white salts with the alkalis, a splendid red salt with mercury, and a yellow one with lead. The salts of iron vary from white and yellow to green and dark brown. Those of copper, a red metal, are of a beautiful blue and green colour, and the anhydrous sulphate is white.

Isomorphism, which appears in a very remarkable manner among the organic compounds, has, under the head of crystallization, already had our attention. There is also a class of bodies which are said to be *isomeric;* that is, to have the same composition, although different in their physical characters. But the idea that bodies exist, which, although of a decidedly different external character, are of exactly the same chemical composition and physical condition, is not tenable; and in nearly all the examples which have been carefully examined, a difference in the aggregate number of atoms, or in the mode in which those atoms have respectively arranged themselves, or that peculiar physical difference designated by the term *allotropy*, has been detected.

Oil of turpentine and oil of lemons have the same composition, each being composed of five equivalents of carbon and four of hydrogen. These substances form, from the striking difference perceptible in their external characters, a good example of isomerism.

The laws of organic chemistry are not, however, the same as those applying to inorganic combinations. Organic chemistry is well defined by Liebig, as the chemistry of compound radicals; and under the influence of vitality, nature produces compounds which have all the properties of simple elements. ([203])

When we reflect upon the conditions which prevail throughout nature, with a few of which only has science made us acquainted, we cannot fail to be struck with the various phases of being which are presented to our observation, and the harmonious system upon which they all appear to depend.

When we discover that bodies are formed of certain determinate atoms, which unite one with another, according to an arithmetical system, to form molecules, which, combining with molecules, observe a similar law, we see at once that all the harmonies of chemical combination — the definite proportions, laws of volume, and the like — are but the necessary consequences of these simple and guiding first principles. In the pursuit of truth, investigators must discover still further arrangements, which, from their perfection, may be compared to the melodious interblending of sweet sounds, and many of the apparently indeterminate combinations will, beyond a doubt, be shown to be as definite as any others. But we cannot reflect upon the fact that these atoms and these molecules are guided in their combinations by impulses, which we can only explain by reference to human passions, as the term *elective affinity* implies, without feeling that an impenetrable mystery of a grand and startling character in its manifestations surrounds each grain of dust which is hurried along upon the wind.

We now, habitually, speak of attraction and repulsion — of the affinity and non-affinity of bodies. We are disposed, from the discovery of the attractive and repelling poles of electrified substances, to regard these powers in all cases as depending upon some electrical state, and we write learnedly upon the laws of these forces. After all, it would be more honest to admit, that we know no more of the secret impulses which regulate the combinations of matter, than did those who satisfied themselves by referring all phenomena of these kinds to sympathies and antipathies: terms which have a poetic meaning, conveying to the mind, with considerable distinctness, the fact, and giving the idea of a feeling — a passion — involving and directing inanimate matter, similar to that which stirs the human heart, and certainly calculated to convey the impression that there is working within all things a living principle, and pointing, indeed, to "the soul of the world." The animated marble of ancient story is far less wonderful than the fact, proved by investigation, that every atom of matter is penetrated by a principle which directs its movements and orders its positions, and involved by an influence which extends, without limits, to all other atoms, and which determines their union, or otherwise.

We have gravitation, drawing all matter to a common centre, and acting from all bodies throughout the wide regions of unmeasured space upon all. We have cohesion, holding the particles of matter enchained, operating only at distances too minute for the mathematician to measure; and we have chemical attraction, different from either of these, working no less mysteriously within absolutely insensible distances, and by the exercise of its occult power, giving determinate and fixed forms to every kind of material creation.

The spiritual beings, which the poet of untutored nature gave to the forest, to the valley, and to the mountain, to the lake, to the river, and to the ocean, working within their secret offices, and moulding for man the beautiful or the sublime, are but the weak creations of a finite mind, although they have for us a charm which all men unconsciously obey, even when they refuse to confess it. They are like the result of the labours of the statuary, who, in his high dreams of love and sublimated beauty, creates from the marble block a figure of the most exquisite moulding which mimics life. It charms us for a season; we gaze and gaze again, and its first charms vanish; it is ever and ever still the same dead heap of chiselled stone. It has not the power of presenting to our wearying eyes the change which life alone enables matter to give; and we admit the excellence of the artist, but we cease to feel at his work. The poetical creations are pleasing, but they never affect the mind in the way in which the poetic realities of nature do. The sylph moistening a lily is a sweet dream; but the thoughts which rise when first we learn that its broad and beautiful dark-green leaves, and its pure and delicate flower, are the results of the alchemy which changes gross particles of matter into symmetric forms, — of a power which is unceasingly at work under the guidance of light, heat, and electrical force, — are, after our incredulity has passed away

—for it is too wonderful for the untutored to believe at once—of an exalting character.

The flower has grown under the impulse of principles which have traversed to it on the beam of solar light, and mingled with its substance. A stone is merely a stone to most men. But within the interstices of the stone, and involving it like an atmosphere, are great and mighty influences, powers which are fearful in their grander operations, and wonderful in their gentler developments. The stone and the flower hold, locked up in their recesses, the three great known forces—light, heat, and electricity: and, in all probability, others of a more exalted nature still, to which these powers are but subordinate agents. Such are the facts of science, which, indeed, draw "sermons from stones," and find "tongues in trees." How weak are the creations of romance, when viewed beside the discoveries of science! One affords matter for meditation, and gives rise to thoughts of a most ennobling character; the other excites for a moment, and leaves the mind vacant or diseased. The former, like the atmosphere, furnishes a constant supply of the most healthful matter; the latter gives an unnatural stimulus, which compels a renewal of the same kind of excitement, to maintain the continuation of its pleasurable sensations.

CHAPTER XII.

CHEMICAL PHENOMENA.

Water — Its Constituents — Oxygen — Hydrogen — Peroxide of Hydrogen — Physical Property of Water — Ice — Sea Water — Chlorine — Muriatic Acid — Iodine — Bromine — Compounds of Hydrogen with Carbon — Combustion — Flame — Safety Lamp — Respiration — Animal Heat — The Atmosphere — Carbonic Acid — Influence of Plants on the Air — Chemical Phenomena of Vegetation — Compounds of Nitrogen — Mineral Kingdom, &c. &c.

Without attempting anything which shall approach even to the character of a sketch of chemical science, we may be allowed, in our search after exalting truths, to select such examples of the results of combination as may serve to elucidate any of the facts connected with natural phenomena. In doing this, by associating our examination with well-known natural objects or conditions, the interpretation afforded by analysis will be more evident, and the operation of the creative forces rendered more striking and familiar, particularly if at the same time we examine such physical conditions as are allied in action, and are sufficiently explanatory of important features.

A large portion of this planet is covered by the waters of the ocean, of lakes and rivers. Water forms the best means of communication between remote parts of the earth. It is in every respect of the utmost importance to the animal and vegetable kingdom; and, indeed, it is indispensable in all the great phenomena of the inorganic world. The peculiarities of saltness or freshness in water are dependent upon its solvent powers. The waters of the ocean are saline from holding dissolved various saline compounds, which are received in part from, and imparted also to, the marine plants. Perfectly pure water is without taste; even the pleasant character of freshly-drawn spring-water is due to the admixture of carbonic acid. It is chemically composed of two volumes of hydrogen gas — the lightest body known, and at the same time highly inflammable — united with one volume of oxygen, which excites combustion, and continues that action, producing heat and light with great energy. By weight, one part of hydrogen is united with eight of oxygen, or in 100 parts of water we find 88.9 oxygen, and 11.1 of hydrogen gas. That two such bodies should unite to furnish the most refreshing beverage, and indeed the only natural drink for man and animals, is one of the extraordinary facts of science. Hydrogen will not support life — we cannot breathe it and live; and oxygen would over-stimulate the organic system, and, producing a kind of combustion, give rise to fever in the animal frame; but, united, they form that drink, for a drop of which the fevered monarch would yield his diadem, and the deprivation of which is one of the most horrid calamities that can be inflicted upon any living or moving thing. Water appears as the antagonist principle to fire, and the ravages of the latter are quenched by the assuaging powers of the former; yet a mixture of oxygen and hydrogen gases, in the exact proportion in which they form water, explodes with the utmost violence on the contact of flame, and, when judiciously arranged, produces the most intense degree of heat known; — such is the remarkable difference between a merely mechanical mixture and a chemical combination.

If we place in a globe, oxygen and hydrogen gases, in the exact proportions in which they combine to form water, they remain without change of state. They appear to mix intimately; and, notwithstanding the difference in the specific gravities of the two gases, the lighter one is diffused through the heavier in a curious manner, agreeably to a law which has an important bearing on the conditions of atmospheric phenomena.(209) The moment, however, that an incandescent body, or the spark from an electric machine, is brought into contact with the mixed gases, they ignite, explode violently, and combine to form water. The discovery of the composition of water was thus synthetically made by Cavendish — its constitution having been previously theoretically announced by Watt.(210)

If, instead of combining oxygen and hydrogen in the proportions in which they form water, we compel the hydrogen to combine with an additional equivalent of oxygen, we have a compound possessing many properties strikingly different from water. This — peroxide of hydrogen, as it is called — is a colourless liquid, less volatile than water, having a metallic taste. It is decomposed at a low temperature, and, at the boiling point, the oxygen escapes from it with such violence, that something like an explosion ensues. All metals, except iron, tin, antimony, and tellurium, have a tendency to decompose this compound, and separate it into oxygen and water. Some metals are oxidized during the decomposition, but gold, silver, platinum, and a few others, still retain their metallic state. If either silver, lead, mercury, gold, platinum, manganese, or cobalt, in their highest states of oxidation, are put into a tube, containing this peroxide of hydrogen, its oxygen is liberated with the rapidity of an explosion, and so much heat is excited, that the tube becomes red hot. These phenomena, to which we have already referred in noticing catalysis, are by no means satisfactorily explained, and the peculiar bleaching properties possessed by the peroxide of hydrogen, sufficiently distinguish it from water. There are few combinations which show more strikingly than these, the difference arising from the chemical union of an additional atom of one element. Similar instances are numerous in the range of chemical science; but scarcely any two exhibit such dissimilar properties. During the ordinary processes of combustion, it has been long known that water is formed by the combination of the hydrogen of the burning body with the oxygen of the air. The recent researches of Schönbein have shown that this peroxide of hydrogen — or, as he calls it, ozone — is produced at the same time, and that it is developed in a great many ways, particularly during electrical changes of the atmosphere. Thus we obtain evidence that this remarkable compound, which was considered as a chemical curiosity merely, is diffused very generally through nature, and produced under a great variety of circumstances. During the excitation of an electrical machine, or the passage of a galvanic current through water by the oxidation of phosphorus, and probably in many similar processes — in particular those of combustion, and we may, therefore, infer also of respiration — this principle is formed. From observations which have been made, it would appear that, during the night, when the activity of plants is not excited by light, they act upon the atmosphere in such a way as to produce this peroxide of hydrogen; and its presence is said to be indicated by its peculiar odour during the early hours of morning. We are not yet acquainted with this body sufficiently to speculate on its uses in nature; without doubt, they are important, perhaps second to those of water only. It is probable that ozone may be the active agent in removing, from the atmosphere, those organic poisons to which many forms of pestilence are traceable; and it is a curious fact, that a low electrical intensity, and a consequent deficiency of atmospheric ozone, marks the prevalence of cholera, and an excess distinguishes the reign of influenza.([211])

Water is one of the most powerful chemical agents, having a most extensive range of affinities, entering directly into the composition of a great many crystallizable bodies and organic compounds. In those cases where it is not combined as water, its elements often exist in the proportions in which water is formed. Gum, starch, and sugar only differ in the proportions in which the elements of water are combined with the carbon.

In saline combinations, and also in many organic forms, we must regard the water as condensed to the solid form; that is, to exist as ice. We well know that, by the abstraction of heat, this condition is produced; but, in chemical combinations, this change must be the result of the mechanical force, exerted by the power of the agency directing affinity.

In the case of water passing from a liquid to a solid state, we have a most beautiful exemplification of the perfection of natural operations. Water conducts heat downwards but very slowly; a mass of ice will remain undissolved but a few inches under water, on the surface of which, ether, or any other inflammable body, is burning. If ice (solid water) swam beneath the surface, the summer sun would scarcely have power to thaw it; and thus our lakes and seas would be gradually converted into solid masses.

All similar bodies contract equally during the process of cooling; and if this applied to water, it has been thought that the result would be the sudden consolidation of the whole mass. A modification of the law has been supposed to take place to suit peculiar circumstances of water. Nature never modifies a law for a particular purpose; we must, therefore, seek to explain the action of the formation of ice, as we know it, by some more rational view.

Water expands by heat, and contracts by cold; consequently, the coldest portions of this body occupy the lower portions of the fluid; but it must be remembered that these parts are warmed by the earth. Ross, however, states that at the depth of 1,000 fathoms the sea has a constant temperature of 39°. Water is at its greatest density at 40° of Fahrenheit's thermometer; in cooling further, it appears to expand, in the same way as if heated;

and, consequently, water, colder than this point, instead of being heavier, is lighter, and floats on the surface of the warmer fluid. It does not seem that any modification of the law is required to account for this phenomenon. Water cooled to 40° still retains its peculiar corpuscular arrangement; but immediately it passes below that temperature, it begins to dispose itself in such a manner that visible crystals may form the moment it reaches 32°. Now, if we conceive the particles of water at 39° to arrange themselves in the manner necessary for the assumption of the solid form, by the particular grouping of molecules in an angular instead of a spheroidal shape, it will be clear, from what we know of the arrangement of crystals of water — ice — that they must occupy a larger space than when the particles are disposed, side by side, in minute spheres. This expansion still goes on increasing, from the same cause, during the formation of ice, so that its specific gravity is less than that of water at any temperature below 40°.

Water, at rest, may be cooled many degrees below the freezing point without becoming solid. This is easily effected in a thin glass flask; but the moment it is agitated, it becomes a firm mass. Here we have another cause aiding in producing crystals of ice on the surface of water, under the influence of the disturbance produced by the wind, which does not extend to any depth.

As oxygen and hydrogen gases enter largely into other chemical compounds beside water, it is important to consider some of the forms of matter into the composition of which these elements enter. To examine this thoroughly, a complete essay on chemical philosophy would be necessary; we must, therefore, be content with referring to a few of the more remarkable instances.

The waters of the ocean are salt; this arises from their holding, in solution, muriate of soda (common culinary salt) and other saline bodies. This muriate of soda is a compound of muriatic acid and soda: muriatic acid is hydrogen, combined with a most remarkable gaseous body, called, from its yellow colour, *chlorine;* and soda, oxygen in union with the metal sodium. Chlorine, in some respects, resembles oxygen; it attacks metallic bodies with great energy; and, in many cases, produces the most vivid incandescence, during the process of combination. It is a powerful bleaching agent, is destructive to animal life, and rapidly changes all organic tissues. There are two other bodies in many respects so similar to chlorine, although the one is at the ordinary temperatures solid, and the other fluid, and which are also discovered in sea-water, or in the plants growing in it, that it is difficult to consider them otherwise than as different forms of the same principle. These are iodine and bromine, and they both unite with hydrogen to form acids. The part which chlorine performs in nature is a great and important one. As muriate of soda, we may trace it in large quantities through the three kingdoms of nature, and the universal employment of salt as a condiment, indicates the importance to the animal economy of the elements composing it. Iodine has been traced through the greater number of marine plants, existing, apparently, as an essential element of their constitution; it has been detected in some mineral springs, and in small quantities in the mineral kingdom.([212]) Bromine is found in sea-water, although in extremely minute quantities, and in a few saline springs; but we have no evidence to show that its uses are important in nature.

Hydrogen, again, unites with carbon in various proportions, producing the most dissimilar compounds. The air evolved from stagnant water, and the fire-damp of the coal mine, are both carburetted hydrogen; and the gas which we employ so advantageously for illumination, is the same, holding an additional quantity of carbon in suspension. Naphtha, and a long list of organic bodies, are composed of these two chemical elements.

These combinations lead us, naturally, to the consideration of the great chemical phenomena of combustion, which involve, indeed, the influences of all the physical powers. By the application of heat, we produce an intense action in a body said to be combustible; it burns, — a chemical action of the most energetic character is in progress, the elements which constitute the combustible body are decomposed, they unite with some other elementary principles, and new compounds are formed. A body burns — it is entirely dissipated, or it leaves a very small quantity of ashes behind unconsumed, but nothing is lost. Its volatile parts have entered into new arrangements, the form of the body is changed, but its constituents are still playing an important purpose in creation.

The ancient notion that fire was an empyreal element, and the Stahlian hypothesis of a phlogistic principle on which all the effects of combustion depended,([213]) have both given way to the philosophy of the unfortunate Lavoisier — which has, indeed, been modified in our own times — who showed that combustion is but the development of heat and light under the influence of chemical combination.

Combustion was, at one period, thought to be always due to the combination of oxygen with the body burning, but research has shown that vivid combustion may be produced where there is no oxygen. The oxidizable metals burn most energetically in chlorine, and some of them in the vapour of iodine and bromine, and many other unions take place with manifestations of incandescence. Supporters of combustion were, until lately, regarded as bodies distinct from those undergoing combustion. For example, hydrogen was regarded as a combustible body, and oxygen as a supporter of combustion. Such an arrangement is a most illogical one, since we may burn oxygen in contact with hydrogen, in the same manner as we burn hydrogen in contact with oxygen; and so in all the other cases, the supporter of combustion may be burnt in an atmosphere formed of the, so called, combustible. The ordinary phenomena of combustion are, however, due to the combination of oxygen with the body burning; therefore every instance of oxidization may be regarded as a condition of combustion, the difference being only one of degree.

Common iron, exposed to air and moisture, rusts, it combines with oxygen. Pure iron, in a state of fine division, unites with oxygen so readily, that it becomes incandescent, and in both cases oxide of iron is formed. This last instance is certainly a case of combustion; but in what does it differ from the first one, except in the intensity of the action? The cases of spontaneous combustion which are continually occurring, are examples of an analogous character to the above. Oxygen is absorbed, it enters more or less quickly, according to atmospheric conditions, into chemical combination, heat is evolved, and eventually, the action continually increasing, true combustion takes place. In this way our cotton-ships, storehouses of flax, piles of oiled-cloth, sawdust, &c., frequently ignite; and to such an influence is to be attributed the destruction of two of our ships of war, a few years since, in Devonport naval arsenal.([214])

In the economic production of heat and light, we have the combination of hydrogen and carbon with the oxygen of common air, forming water and carbonic acid. In our domestic fires we employ coal, which is essentially a compound of carbon and hydrogen, and some matters which must be regarded as impurities; the taper, whether of wax or tallow, is made up of the same bodies, differing only in their combining proportions, and, like coal gas, these burn as carburetted hydrogen. All these bodies are very inflammable, having a tendency to combine energetically with oxygen at a certain elevation of temperature.

We are at a loss to know how heat can cause the combination of those bodies. Sir Humphry Davy has shown that hydrogen will not burn, nor a mixture of it with oxygen explode, unless directly influenced by a body heated so as to *emit light.*(215) May we not, therefore, conclude that the chemical action exhibited in a burning body is a development of some latent force, with which we are unacquainted, produced by the absorption of light; — that a repulsive action at first takes place, by which the hydrogen and carbon are separated from each other; — and that in the nascent state they are seized by the oxygen, and again compelled, though in the new forms of water and carbonic acid, to resume their chains of combining affinity.

Every equivalent of carbon and of hydrogen in the burning body unites with two equivalents of oxygen, in strict conformity with the laws of combination. The flame of hydrogen, if pure, gives scarcely any light, but combined with the solid particles of carbon, it increases in brightness. The most brilliant of the illuminating gases is the olefiant gas, produced by the decomposition of alcohol, and it appears to be hydrogen charged with carbon to the point of saturation. Flame is a cone of heated vapour, becoming incandescent at the points of contact with the air, a mere superficial film only being luminous. It is evident that all the particles of the gas are in a state of very active repulsion over the surface, since flame will not pass through wire gauze of moderate fineness. Upon this discovery is founded the inimitable safety-lamp of Davy, by means of which the explosive gases of a mine are harmlessly ignited within a cage of wire gauze. This effect has been attributed to a cooling influence of the metal; but, since the wires may be brought to a degree of heat but little below redness without igniting the fire-damp, this does not appear to be the cause. It appears to present an example exactly the converse of that already stated with reference to the spheroidal state of water, and it affords additional evidence that the condition of bodies at high temperatures is subject to important physical changes.

The researches which led to the safety-lamp may be regarded as among the most complete examples of correct inductive experiment in the range of English science, and the result is certainly one of the proudest achievements of physico-chemical research. By merely enveloping the flame of a lamp with a metallic gauze, the labourer in the recesses of the gloomy mine may feel himself secure from that outpouring current of inflammable gas, which has been so often the minister of death; he may walk unharmed through the explosive atmosphere, and examine the intensity of its power, as it is wasted in trifling efforts within the little cage he carries. Accidents have been attributed to the "Davy," as the lamp is called among the colliers; but they may in most cases be traced to carelessness on the part of those whose duty it has been to examine the lamps, or to the recklessness of the miners themselves.

That curious metal, platinum, and also palladium, possesses a property of maintaining a slow combustion, which has been rendered available by the discoverer of the safety-lamp to a very important purpose. If we take a coil of platinum wire, and, having made it red-hot, plunge it into an explosive atmosphere of carburetted hydrogen and common air, it continues to glow with considerable brightness, producing, by this very peculiar influence, a recombination of the gases, which is discovered by the escape of pungent acid vapours. Over the little flame of the safety-lamp a coil of platinum is suspended, and it is thus kept constantly at a red heat. If the miner becomes accidentally enveloped in an atmosphere of fire-damp, although the flame of his lamp may be extinguished, the wire continues to glow with sufficient brightness to light him from his danger, through the dark winding passages which have been worked in the bed of fossil fuel.

It is thus that the discoveries of science, although they may appear of an abstract character, constantly, sooner or later, are applied to uses by which some branch of human labour is assisted, the necessities of man's condition relieved, and the amenities of life advanced.

The respiration of animals is an instance of the same kind of chemical phenomena as we discover in ordinary combustion. In the lungs the blood becomes charged with oxygen, derived from the atmospheric air, with which it passes through the system, performing its important offices, and the blood is returned to the lungs with the carbonic acid, formed by the separation of carbon from the body, which is thrown off at every expiration. It will be quite evident that this process is similar to that of ordinary combustion. In man or animals, as in the burning taper,—which is aptly enough employed by poets as the symbol of life,—we have hydrogen and carbon, with some nitrogen superadded; the hydrogen and oxygen form water under the action of the vital forces; the carbon with oxygen produces carbonic acid, and, by a curious process, the nitrogen and hydrogen also combine to form ammonia.([216])

All the carbon which is taken into the animal economy passes, in the process of time, again into the atmosphere, in combination with oxygen, this being effected in the body, under the *catalytic* power of tissues, immediately influenced by the excitation of nervous forces, which are the direct manifestations of vital energy. The quantity of carbonic acid thus given out to the air is capable of calculation, with only a small amount of error. It appears that upwards of fifty ounces of carbonic acid must be given off from the body of a healthy man in twenty-four hours. On the lowest calculation, the population of London must add to the atmosphere daily 4,500,000 pounds of carbonic acid. It must also be remembered that in every process for artificial illumination, and in all the operations of the manufactures in which fire is used, and also in our arrangements to secure domestic comfort, immense quantities of this gas are formed. We may, indeed, fairly estimate the amount, if we ascertain the quantity of wood and coal consumed, of all the carbon which combines with oxygen while burning, and escapes into the air, either as carbonic acid or carbonic oxide. The former gas, the same as that which accumulates in deep wells and in brewers' vats, is highly destructive to life, producing very distressing symptoms, even when mixed with atmospheric air, in but slight excess over what it commonly contains. The oppressive atmosphere of crowded assemblies, is in a great measure due to the increased proportion of carbonic acid it contains. It will be evident to every one, that, unless some provision was made for removing this deleterious gas from the atmosphere as speedily as it formed, consequences of the most injurious character to the animal races would ensue. It is found, however, that the quantity in the atmosphere is almost constantly about one per cent. The peculiar properties of carbonic acid, in part, insure its speedy removal. It is among the heaviest of gaseous bodies, and it is readily absorbed by water; consequently, floating within a short distance from the surface of the earth, a large quantity is dissolved by the waters spread over it. A large portion is removed by the vegetable kingdom; indeed, the whole of that produced by animals, and by the processes of combustion, eventually becomes part of the vegetable world, being absorbed with water by the roots, and separated from the air by the peculiar functions of the leaves. However, the property of the diffusion of gases explains the rapid mixing of this heavy gas with the lighter atmospheric fluid.

The leaves of plants may be regarded as performing similar offices to the lungs of animals. They are the breathing organs. In the animal economy a certain quantity of carbon is necessarily retained, in combination with nitrogen and other elements, to form muscle; but this constantly undergoing change; — the entire system being renewed within a comparatively limited period. The conditions with plants are somewhat different. For instance, the carbon is fixed in a tree, and remains as woody fibre until it decays, even though the life of the plant may extend over many centuries.

Animals, then, are constantly supplying carbonic acid; plants are as constantly feeding on it; thus is the balance for ever maintained between the two kingdoms. Another condition is, however, required to maintain for the uses of men and animals the necessary supply of oxygen gas. This is effected by one of those wonderful operations of nature's chemistry which must strike every reflecting mind with admiration. During the night plants breathe carbonic acid; but there is a condition of repose prevailing then in their functions, and much of it passes off unchanged. With the first gleam of the morning sun the dormant organs of the plant are awakened into full action; they decompose this carbonic acid, secrete the carbon, to form the rings of wood which constitute so large a part of their structure, and pour out pure oxygen gas to the air. The plant is, therefore, an essential element in the conditions necessary for the support of animal life. It must necessarily follow, that the inhabitants of the tropics do not produce so much carbonic acid as those who dwell in colder regions. In the first place, their habits of life are different, and they are not under the necessity of maintaining animal heat by the use of artificial combustion, as are the people of colder climes. The vegetation of the regions of the tropics is much more luxuriant than that of the temperate and arctic zones.

Hence an additional supply of carbonic acid is required between the torrid zones, and a less quantity is produced by its animals. These cases are all met by the great aërial movements. A current of warmed air, rich in oxygen, moves from the equator towards the poles, whilst the cooler air, charged with the excess of carbonic acid, sets in a constant stream towards the equator. By this means the most perfect equalization of the atmospheric conditions is preserved.

The carbonic acid poured out from the thousand mouths of our fiery furnaces, — produced during the laborious toil of the hard-working artizan, — and exhaled from every populous town of this our island home—is borne away by the prevailing aërial currents to find its place in the pines of the Pacific Islands, the spice-trees of the Eastern Archipelago, and the cinchonas of Southern America. The plants of the valley of the Caucasus, and those which flourish amongst the Himalayas, equally with the less luxuriant vegetation of our temperate climes, are directly dependent upon man and the lower animals for their supply of food.

If all plants were removed from the earth, animals could not exist. How would it be if the animal kingdom was annihilated ? — would it be possible for vegetation to continue ? This question is not quite so easily answered, for it has been supposed that during the epoch of the coal formation a luxuriant vegetation must have gone on over the earth's surface, and the evidences of the existence of animal life during that period are but few. It is supposed that the air was then charged with carbonic acid, and that the calamites, lepidodendra, and sigillaria were employed to remove it, and fit the earth for the oxygen-breathing races. The evidence upon these points is by no means satisfactory; and although at one time quite disposed to acquiesce in a conjecture which appears to account so beautifully for the observed geological phenomena of carboniferous periods, we do not regard the necessities for such a condition of the atmosphere as clearly made out.([217])

In all probability the same mutual dependence, which now exists between the animal and vegetable kingdoms, existed from the beginning of time, and will continue to do so under varying circumstances through the countless ages of the earth's duration.

There is yet another very important chain of circumstances which binds these two great kingdoms together. This is the chain of the animal necessities. A large number of races feed directly upon vegetables; herbs and fruits are the only things from which they gain those elements required to restore the waste of their systems.

These herbivorous animals, which must necessarily form fat and muscle from the elements of their vegetable diet, are preyed on by the carnivorous races; and from these the carbon is again restored to the vegetable world. Sweep off from the earth the food of the herbivora, they must necessarily very soon perish, and, with their dissolution, the destruction of the carnivora is certainly insured. To illustrate this, on a small scale, it may be mentioned that around the coasts of Cornwall, pilchards were formerly caught in very great abundance, in the shallow water within coves, where these fish are now but rarely seen. From the investigations of the Messrs. Couch, whose very accurate observations on the Cornish fauna have placed both father and son amongst the most eminent of British naturalists,([218]) it appears that the absence of these fish is to be attributed entirely to the practice of the farmers, who cut the sea-weed from the rocks for the purpose of manuring their lands. By this they destroy all the small crustacea inhabiting these immature marine forests, feeding on the algæ, and as these, the principal food of the pilchards, have perished, they seek for a substitute in more favourable situations. Mr. Darwin remarks that, if the immense sea-weeds of the Southern Ocean were removed by any cause, the whole fauna of these seas would be changed.

We have seen that animals and vegetables are composed principally of four elementary principles, oxygen, hydrogen, nitrogen, and carbon. We have examined the remarkable manner in which they pass from one condition — from one kingdom of nature — into another. The animal, perishing and dwindling by decomposition into the most simple forms of matter, mingling with the atmosphere as mere gas, gradually becomes part of the growing plant, and by like changes vegetable organism progresses onward to form a portion of the animal structure.

A plant exposed to the action of natural or artificial decomposition passes into air, leaving but a few grains of solid matter behind it. An animal, in like manner, is gradually resolved into "thin air." Muscle, and blood, and bones, having undergone the change, are found to have escaped as gases, leaving only "a pinch of dust," which belongs to the more stable mineral world. Our dependency on the atmosphere is therefore evident. We derive our substance from it — we are, after death, resolved again into it. We are really but fleeting shadows. Animal and vegetable forms are little more than consolidated masses of the atmosphere. The sublime creations of the most gifted bard cannot rival the beauty of this, the highest and the truest poetry of science. Man has divined such changes by the unaided powers of reason, arguing from the phenomena which science reveals in unceasing action around him. The Grecian sage's doubts of his own identity, was only an extension of a great truth beyond the limits of our reason. Romance and superstition resolve the spiritual man into a visible form of extreme ethereality in the spectral creations, "clothed in their own horror," by which their reigns have been perpetuated.

When Shakspeare made his charming Ariel sing —

> "Full fathom five thy father lies,
> Of his bones are coral made,
> Those are pearls that were his eyes:
> Nothing of him that doth fade,
> But doth suffer a sea change
> Into something rich and strange;"

he painted, with considerable correctness, the chemical changes, by which decomposing animal matter is replaced by a siliceous or calcareous formation.

The mysteries of flowers have ever been the charm of the poet's song. Imagination has invested them with a magic influence, and fancy has almost regarded them as spiritual things. In contemplating their surpassing loveliness, the mind of every observer is improved, and the sentiments which they inspire, by their mere external elegance, are great and good. But in examining the real mysteries of their conditions, their physical phenomena, the relations in which they stand to the animal world, "stealing and giving odours" in the marvellous interchange of carbonic acid and ammonia for the soul-inspiring oxygen — all speaking of the powers of some unseen, indwelling principle, directed by a Supreme Ruler — the philosopher finds subjects for deep and soul-trying contemplation. Such studies lift the mind into the truly sublime of nature. The poet's dream is the dim reflection of a distant star: the philosopher's revelation is a strong telescopic examination of its features. One is the mere echo of the remote whisper of Nature's voice in the dim twilight; the other is the swelling music of the harp of Memnon, awakened by the sun of truth, newly risen from the night of ignorance.

To return from our long, but somewhat natural, disgression, to a consideration of the chemical phenomena connected with the atmosphere, and its curious and important element, nitrogen, we must first consider the evidence we have of the condition of the air itself.

The mean pressure exerted upon the surface of the earth, as indicated by the barometer, is equal to a column of mercury, thirty inches high; that is, the column of air pressing upon the open end of a bent tube filled with mercury, exactly balances that quantity, which represents a pressure of fifteen pounds upon every square inch of surface. This pressure, it must be remembered, is the compound weight of the gaseous envelope, and the elastic force of the aqueous vapour contained in it.([219]) If the atmosphere were of uniform condition, its height, as inferred from the barometer, would be about five miles and a half. The density of the air, however, diminishes with the pressure upon it, so that at the height of 11,556 feet the atmosphere is of half density; or one volume of air, as taken at the surface of the earth, is expanded into two. Thus the weight is continually diminishing; but this is regularly opposed by the decreasing temperature, which is at the rate of about one degree for every 352 feet of ascent, although in all probability it is less rapid at great distances from the earth.

It has been calculated from certain phenomena of refraction, that our atmosphere must extend to about forty miles from the surface of the earth. It may, in a state of extreme tenuity, extend still further; but it is probable that the intense cold produced by rarefaction, sets limit to any extension much beyond this.

The uses of the atmosphere are many. It is the medium for regulating the dispersion of watery vapours over the earth. If there was no atmosphere, and that, as now, the equatorial climes were hot and the poles cold, evaporation would be continually going on at the equator, and condensation in the colder regions. The sky of the tropical climes would be perpetually cloudless, whilst in the temperate and artic zones we should have constant rain and snow. By having a gaseous atmosphere, a more uniform state of things is produced, the vapours evaporated from the earth become intimately mixed with the air, and are borne by it over large tracts of country, and only precipitated when they enter some stratum much colder than that which involves them. There are opposite tendencies in an atmosphere of air and one of vapour. The air circulates from the colder to the warmer parts, and the vapour from the warmer to the colder regions; and as the currents of the air, from the distribution of land and sea — the land from its low conducting power being more quickly heated than the sea — are very complicated, and as some force is employed in keeping the vapour suspended in the air, water is less suddenly deposited on the earth than it would have been, had not these tendencies of the air and its hygrometric peculiarities been as we find them.

The blue colour of the sky, which is so much more agreeable to the eye than either red or yellow, is due to a tendency of the mixed gas and vapour to reflect the blue rays rather than red or yellow. The white light which falls upon the surface of the earth, without absorption or decomposition in its passage from the sun, is partially absorbed by, and in part reflected back from, the earth. The reflected rays pass with tolerable freedom through this transparent medium, but a portion of the blue rays are interrupted and rendered visible to us. That it is reflected light, is proved by the fact of its being in a polarized state.([29]) Clouds of vapour reflect to us again, not isolated rays, but the undecomposed beam, and consequently they appear white as snow to our vision.

The golden glories of sunset, — when, "like a dying dolphin," heaven puts on the most gorgeous hues, which are continually changing, — depend entirely upon the quantity of watery vapour which is mixed with the air, and its state of condensation. It has been observed, that steam at night, issuing into the atmosphere under a pressure of twenty or thirty pounds to the square inch, transmits and reflects orange-red light. This we may, therefore, conclude to be the property of such a condition of mixed vapour and air, as prevails when the rising or the setting sun is shedding over the eastern or the western horizon the glory of its coloured rays. ([221])

Thus, science points out to us the important uses of the air. We learn that life and combustion are entirely dependent on it, and that it is made the means for securing greater constancy in the climates of the earth than could otherwise be obtained. The facts already dwelt upon are sufficient to convince every thinking mind, that the beautiful system of order which is displayed in the composition of the atmosphere, in which the all-exciting element, oxygen, is subdued to a tranquil state by another element, nitrogen, (which, we shall have presently to show, is itself, under certain conditions, one of the most energetic agents with which we are acquainted,) indicates a Supreme power, omniscient in the adaptation of things to an especial end. Oxygen and nitrogen are here *mixed* for the benefit of man; man *unites* them by the aid of powers with which he is gifted, and the consequences are of a fatal kind. The principles which the great chemist of nature renders mild are transformed into sources of evil by the chemist of art.

Beyond all this, the atmosphere produces effects on light which add infinitely to the beauty of the world. Were there no atmosphere, we should only see those objects upon which the sun's rays directly fell, or from which they were reflected. A ray falling through a small hole into a dark room, illuminating one object, which reflects some light upon another, is an apt illustration of the effect of light upon the earth, if it existed without its enveloping atmosphere. By the dispersive powers of this medium, sunlight is converted into daylight; and instead of unbearable parallel rays illuminating brilliantly, and scorching up with heat those parts upon which they directly fall, leaving all other parts in the darkness of night, we enjoy the blessings of a diffusion of its rays, and experience the beauties of soft shades and slowly-deepening shadows. Without an atmosphere, the sun of the morning would burst upon us with unbearable brilliancy, and leave us suddenly, at the close of day, at once in utter darkness. With an atmosphere, we have the twilight with all its tempered loveliness, — a " time for poets made."

In chemical character, atmospheric air is composed of twenty one volumes of oxygen, and seventy-nine volumes of nitrogen: or one hundred grains of air consist of 23.1 grains of the former, and 76.9 grains of the latter. Whether the air is taken from the greatest depths or the most exalted heights to which man has ever reached, an invariable proportion of the gases is maintained. The air of Chimborazo, of the arid plains of Egypt, of the pestilential delta of the Niger, or even of the infected atmosphere of an hospital, all give the same proportions of these two gases, as we find existing on the healthful hills of Devonshire, or in the air of the city of London. This constancy in constitution leads to the supposition that the oxygen and nitrogen are chemically combined; but many eminent philosophers have contended that they are merely mechanically mixed; and they have shown that some peculiar properties prevail amongst gaseous bodies, which very fully explain the equal admixture of two gases, the specific gravities of which are different. This is particularly exemplified in the case of carbonic acid, of which gas one per cent. can be detected in all regions of the air to which the investigations of man have reached. This gas, although so heavy, is, by the law of diffusion, mixed with great uniformity throughout the mass.(22) Every exhalation from the earth, of course, passes into the air; but these are generally either so light that they are carried into the upper regions, and there perform their parts in the meteorological phenomena, or they are otherwise very readily absorbed by water or growing plants, and thus is the atmosphere preserved in a state of purity for the uses of animals. Again, the quantity of oxygen contained in the air, and its very peculiar character, insures the oxidation of all the volatile organic matters which are constantly passing off, — as the odoriferous principles of plants, the miasmata of swamps, and the products of animal putrefaction; these are rapidly converted into water, carbonic acid, or nitric acid, and quickly enter into new and harmless combinations. The elements of contagion we are unacquainted with; but since the attention of inquirers has been of late directed to this important and delicate subject, some light may possibly be thrown upon it before long.

Nothing shows more strikingly the admirable adaptation of all things for their intended uses than the atmosphere. In it we find the source of life and health; and chemistry teaches us, most indisputably, that it is composed of certain proportions of oxygen and nitrogen gases; and experience informs us that it is on the oxygen that we are dependent for all that we enjoy. So beautifully is the atomic or molecular constitution ordered, that it is impossible to produce any change in the air without rendering it injurious to the animal and vegetable economy. It might be thought, from the well-known exhilarating character of oxygen gas, that, if a larger quantity existed in the atmosphere than that which we find there, the enjoyments of life would be of a more exciting kind; but the consequences of any increase would be exceedingly injurious; and, by quickening all the processes of life to an unnatural extent, the animal fabric would soon decay: excited into fever, it would be destroyed by its own fires.

Chemistry has made us acquainted with six other compounds of oxygen and nitrogen, neither of them fitted for the purposes of vitality, of which the following are the most remarkable: —

Nitrous oxide, or the celebrated laughing gas, which contains two volumes of nitrogen to one of oxygen, would prove more destructive than even pure oxygen, from the delirious intoxication which it produces.

Nitric oxide is composed, according to Davy, of two volumes of nitrogen and two of oxygen. It is of so irritating a nature that the glottis contracts spasmodically, when any attempt is made to breathe it; and the moment it escapes into the air, it combines with more oxygen, and forms the deep red fumes of nitrous acid.

Nitrous acid, and the peroxide of nitrogen, each contains an additional proportion of oxygen, and they are still more destructive to all organization.

Nitric acid contains five volumes of oxygen united to two of nitrogen; and the well-known destructive properties of aqua fortis it is unnecessary to describe.

The atmosphere, and these chemically active compounds, contain the same elements, but their mode of combining is different; and what is, in the one case, poisonous to the highest degree, is, in the other, rendered salubrious and essential to all organized beings.

Nitrogen gas may be regarded in the light of a diluent to the oxygen. In its pure state it is only characterized by its negative properties. It will not burn, or act as a supporter of combustion. Animals speedily perish if confined in it; but they die rather through the absence of oxygen than from any poisonous property of this gas. Yet, in combination, we find nitrogen exhibiting powers of a most energetic character. In addition to the fulminating compounds, and the explosive substances already named, which are among the most remarkable instances of unstable affinity with which we are acquainted, we have also the well-known pungent body ammonia. From the analogous nature of this volatile compound, and the fixed alkalies, soda and potash, it was inferred that it must, like them, be an oxide of a metallic base. Davy exposed ammonia to the action of potassium, and to the influence of the voltaic flame produced from 2,000 double plates, without at all changing it character. From its slight tendency to combination, and from its being found abundantly in the organs of animals feeding on substances that do not contain it, it is, however, probably a compound body. A phenomenon of an obscure and mysterious character is presented in the formation of the "ammoniacal amalgam," as it is called.

Mercury, being mixed with an ammoniacal salt, is exposed to powerful galvanic action; and a compound, maintaining its metallic appearance, but of considerable lightness and very porous, presents itself.[23] This preparation has been carefully examined by Davy, Berzelius, and others. It is always resolved into ammonia and mercury; and, although the latter chemist is strongly inclined to regard it as affording evidence of the compound nature of nitrogen,—and he has, indeed, proposed the name *nitricum*, for its hypothetical base,—yet, to the present time, we have no satisfactory explanation of this metallization of ammonia.

No attempt will be made to describe the various elementary substances which come under the class of metallic bodies, much less to enumerate their combinations. Many of the metals, as silver and copper, are found sometimes in a native state, or nearly pure; but, for the most part, they exist, in nature, in combination with oxygen or sulphur — gold furnishes a remarkable exception. They are ordinarily found combined with other bodies, as oxidized carbon, phosphorus, chlorine, &c.; but these cases are by no means so common. Those substances called metals are generally found embedded in the rocks, or deposited in fissures formed through them; but it is one of the great discoveries of modern science, that those rocks themselves are metallic oxides. With metals, we generally associate the idea of great density; but potassium and sodium, the metallic bases of potash and soda, are lighter than water, and they, consequently, float upon that fluid. We learn, therefore, from the researches of science, that the crust of this earth is composed entirely of metals, combined with gaseous elements; and there is reason for believing that one, or perhaps two, of the gases we have already named, are also of a metallic character. Strange as it may appear, there is nothing, as will be seen on attentive consideration, irrational in this idea. Many of the metals proper, under the influence of such heat as we can, by artificial means, command, are dissipated in vapour, and may be maintained in this state perfectly invisible. Indeed, the transparent space above the surface of the mercury in the tube of a barometer, known as the Torricellian vacuum, is filled with the vapour of mercury. There is, therefore, no reason why nitrogen, or even hydrogen, should not be metallic molecules kept by the force of the repulsive powers of heat, or some other influence, at a great distance from each other. The peculiar manner in which nitrogen unites with mercury, and the property which hydrogen possesses of combining with antimony, zinc, arsenic, potassium, sodium, and possibly other metals, besides its union with sulphur and carbon, — in all which cases there is no such change of character as occurs when they combine with oxygen — appear to indicate bodies which, chemically, are not very dissimilar to those metals themselves, although, physically, they have not the most remote resemblance.

"We know nothing," says Davy, "of the true elements

belonging to nature; but so far as we can reason from the relations of the properties of matter, hydrogen is the substance which approaches nearest to what the elements may be supposed to be. It has energetic powers of combination, its parts are highly repulsive as to each other, and attractive of the particles of other matter; it enters into combination in a quantity very much smaller than any other substance, and in this respect it is approached by no known body."([224])

Many of the elements are common to the three kingdoms of nature: most of those found in one condition of organization are found in another. In the vegetable class we find carbon combining with oxygen, and hydrogen, and an inferior quantity of nitrogen. The carbonates are an abundant mineral class. These elements, also, constitute the substance of animals, the proportion of nitrogen being in them much larger. If one element, more than another, belongs especially to the animal economy, it is phosphorus, although this is not wanting in the vegetable world; and it is not uncommon in the mineral. Sulphur is common to the three classes: it is abundant in the mineral kingdom, being one of the products of volcanic action; it is united with the metals, forming sulphurets; it combines with the metallic bases of lime and other earths, and is found in our rocks in the state of sulphuric acid or oxidized sulphur. In the vegetable kingdom we discover sulphur in all plants of the onion kind, in the mustard, and some others; and it enters into the composition of vegetable albumen, and appears always combined with albumen, fibrine, and caseine in the animal economy.

Chlorine is found most abundantly in combination with sodium, as common salt: in this state, in particular, we may trace it from the depths of the earth, its waters, and its rocks, to the plants and animals of the surface. Iodine is most abundant in marine plants; but it has been found in the mineral world.

Bromine is known to us as a product of certain saline waters, and a few specimens of natural bromide of silver have been examined. Fluorine, the base of the acid which, combining with lime, forms fluor-spar, is found to exist to some considerable extent in the bones of animals. It must not be forgotten that the earths enter into the composition of the more solid parts of plants and animals. Silica, or the earth of flints, is met with in beautiful transparent crystals, in the depths of the mine; in all rocks and soils we find it; in the bark of many plants, particularly the grasses, it is discovered, forming the hard supporting cuticle of the stalk of wheat, the Dutch rush, the sugar-cane, the bamboo, and many others. Lime is one of the principal constituents of animal bone and shells, and it is found in nearly all vegetables.

It is thus that we find the same elementary principle presenting itself in every form of matter, under the most Protean shapes. Numerous phenomena of even a more striking character than those selected, are exhibited in every department of chemistry; but within the limits of this essay it is impracticable to speak of any beyond those which directly explain natural phenomena.

The chemical elements, which actually exist in nature as simple bodies, are probably but few. Most of the known gases, and sulphur, phosphorus, and the metals, are in all probability compounds of some ethereal ultimate principles; and with the advance of science we may fairly hope to discover the means of reducing some of them to a yet more simple state. The speculations of men, through all ages, have leaned towards this idea, as is shown by the theory of the four elements of the ancients, the three of the alchemists, and the refined speculations of Newton and Boscovich. All experimental inquiry points towards a similar conclusion. It is true we have no direct evidence of any elementary atom actually undergoing a change of state ; but when we regard the variations produced by electrical influence, and consider the phenomena of allotropism, it will be difficult to come to any other conclusion, than that the particles of matter known to us as ultimate are capable of change, and consequently must be far removed from positively simple bodies, since the real elementary atom, possessing fixed properties, cannot be supposed capable of undergoing any transmutation.

It will now be evident that in all chemical phenomena we have the combined exercise of the great physical forces, and evidences of some powers which are, as yet, shrouded in the mystery of our ignorance. The formation of minerals within the clefts of the rocks, the decomposition of metallic lodes, the germination of seeds, the growth of the plant, the development of its fruit, and its ultimate decay, the secret processes of animal life, assimilation, digestion, and respiration, and all the changes of external form which take place around us, are the result of the exercise of that principle which we call chemical.

By chemical action plants take from the atmosphere the elements of their growth ; these they yield to animals, and from these they are again returned to the air. The viewless atmosphere is gradually formed into an organized being, which as gradually is again resolved into the thin air. The changes of the mineral world are of an analogous character ; but we cannot trace them so clearly in all their phenomena.

An eternal round of chemical action is displayed in nature. Life and death are but two phases of its influences. Growth and decay are equally the result of its power.

CHAPTER XIII.

TIME.—GEOLOGICAL PHENOMENA.

Time, an element in Nature's Operations — Geological Science — Its Facts and Inferences — Nebular Hypothesis applied — Primary Formations — Plutonic and Metamorphic Rocks — Transition Series — Palæozoic Rocks — Commencement of Organic Arrangements — Existence of Phosphoric Acid in Plutonic Rocks — Fossil Remains — Coal Formation — Sand-stones — Tertiary Formations — Eocene, Miocene, and Pliocene Formations — Progressive changes now apparent — General Conclusions — Physics applied in Explanation.

The influence of time, as an element, in producing certain structural arrangements, by modifying the operations of physical force, under whatever form it may be exerted, has scarcely been sufficiently attended to in the examination of cosmical phenomena. Every particle of matter is, as it were, suspended between the agencies to which we have been directing our attention. Under the influences of the physical powers, sometimes exerted in common, but often with a great preponderance in favour of one of them, every accumulated heap of mud or sand is slowly cohering, and assuming the form of a rock possessing certain distinguishing features, as it regards lamination, cleavage, &c.

The minute particles of matter are necessarily but slightly influenced by the physical forces: their action, in accordance with the laws which determine physical condition, is manifested in an exceedingly modified degree. But in all the operations of nature, what is deficient in power is made up in time, and effects are produced during myriads of ages, by powers far too weak to give satisfactory results by any experiments which might be extended even over a century.

If, with the eye of a geologist, we take but a cursory glance over the earth, we shall discover that countless ages must have passed during the progress of this planet to its present state. This is a fact written by the finger of nature, in unmistakeable characters, upon the mighty tablets of her mountains.

The superficial crust of the earth, — by which is meant only that film, — compared with its diameter, — which is represented by a few miles in depth — is composed of distinct mineral masses, exhibiting peculiar physical conditions and a certain order of arrangement. These rocks appear to have resulted from two dissimilar causes; in the one class the action of heat is evident, and in the other we have an aqueous origin indicated by peculiarities of formation.

There are few branches of science which admit of speculation to the extent to which we find it carried in geology. The consequence of this is shown in the popular character of the science. A few observations are made over a limited area, and certain structural conditions are ascertained, and at once the mind, "fancy free," penetrates the profound depths of the earth, and imagination, having "ample room and verge enough," creates causes by which every effect is to be interpreted. Such students, generally ignorant of the first principles of physics, knowing little of mineralogy, and less of chemistry, to say nothing of palæontology, having none of the requisites for an observer, boldly assume premises which are untenable, and think they have explained a phenomenon, — given to the world a truth, — when they have merely promulgated an unsubstantiated speculation, which may have occasional marks of ingenuity, and but little else.

The carefully made observations of those who, with unwearying industry, have traversed hill and valley, marked and measured the various characters, thicknesses, inclinations, and positions of rocks; who have watched the influences of heat in changing, of water in wearing, and the results of precipitation in forming strata; who have traced the mechanical effects of earthquake strugglings and of volcanic eruptions, and, reasoning from an immense mass of accumulated facts, deduced certain general conclusions, — are, however, of a totally different character; and it is such observers as these who induced Herschel to say truly, that "geology, in the magnitude and sublimity of the objects of which it treats, undoubtedly ranks, in the scale of the sciences, next to astronomy." ([225])

The origin of this planet is involved in great obscurity, which the powers of the most gifted are unable to penetrate. It stands the work of an Almighty and Eternal mind, the beginning of which we cannot comprehend, nor can we define the period of its termination.

It may, probably, be safe to speculate that there was a time when this globe consisted of only one homogeneous stratum. Whether this remains, — whether, in our plutonic rocks, our granites, or our porphyries, we have any indications of the primitive state of the world, or whether numerous changes took place before even our unstratified formations had birth, are questions we cannot answer. The geologist looks back into the vista of time, and reckons, by phenomena, the progress of the world's mutations. The stratified formations may have occupied thousands of ages; but before these were, during a period extending over countless thousands, the unstratified rocks may have been variously metamorphosed. It matters not whether we admit the nebular hypothesis or not, a time must have been when all these bodies which now form the mass of this globe existed in the most simple state. We have already shown that very remarkable changes in external character and in chemical relations are induced, in the same simple element, by its having been exposed to pecular and different conditions; and already have we speculated on the probability that the advance of science will enable us to reduce the numerous elements we now reckon, to two or three. It is, therefore, by no means an irrational thought to suppose, that at the beginning a mighty mass of matter, in the most attenuated state, floated through space, and was gradually, under the influence of gravitation, of cohesive force, and of chemical aggregation, moulded into the form of a sphere. Ascending to the utmost refinement of physics, we may suppose that this mass was one of uniform character, and that it became in dissimilar parts — its surfaces and toward its centre — differently constituted, under the influences of the same powers which we now find producing, out of the same body, charcoal and the diamond, and creating the multitudinous forms of organized creations. These conditions being established, and carried to an extent of which, as yet, science has afforded us no evidence, chemical intermixture may have taken place, and a new series of compounds have been formed, which, by again combining, gave rise to another and more complex class of bodies.

The foundation of the superficial crust of the earth appears to be formed of a class of rocks which have resulted from the slow cooling of an immense mass of heated matter. These rocks have been called *igneous;* but are now more generally termed *Plutonic*, (such as granites, syenites, &c.) Immediately above these, we find rocks which have resulted by deposition from water. These masses, having been exposed to the action of the heat below, have been considerably changed in their character, and hence they are often called *metamorphic;* but metamorphic rocks may, however, be of any age. The rocks formerly termed the *transition* series — from their forming the connecting link between the earlier formations — are now, from the circumstance of their being fossiliferous, classed under the general term of palæozoic rocks, to distinguish them from the rocks in which no organic remains have been found. Above these are found the secondary strata, and, still more recently produced, we have a class now usually denominated the tertiary formations. "Eternal as the hills" is a poetic expression, implying a long duration; but these must, from the nature of things, eventually pass away. The period of time necessary for the disintegration of a granite hill is vastly beyond the powers of computation, according to our conception of the ordinary bounds of finite things. But a consideration of the results of a few years,—under the influence of the atmosphere and the rains,—as shown in quantity of solid matter carried off by the rivers, and deposited at their mouths, will tend to carry conviction to every mind, that a degrading process is forever in action on the surface of the earth. The earth itself may be eternal, but the surface is continually undergoing mutation, from various causes, many of which we must briefly consider. ([226])

In regarding geological phenomena, the absence of any fossil remains has often been supposed to indicate a period previous to any organic formations. That the inorganic constituents of matter are of prior origin to the organic combinations is a speculation which must be cautiously received. The supposed evidences in favour of such an assumption are in some respects doubtful; and we can well understand that changes may have been induced in the earlier rock formations, by heat or by other powers, quite sufficient to destroy all traces of organized forms. It was long thought that phosphoric acid was not to be detected in rocks which are regarded as of igneous origin; and since this acid is peculiarly a constituent of organic bodies, this has been adduced as a proof that the plutonic rocks must have existed previously to the appearance of vegetable or animal life upon the surface of the globe. The researches of modern chemists have, however, shown that phosphoric acid is to be found in formations of granitic origin, in porphyry, basalt, and hornblende rocks.(227) If, therefore, we are to regard this substance as of organic origin, the rational inference is against the speculation.

Without attempting to enter into any account of the apparent progress of life over the earth, it appears desirable that some description should be given of the kinds of plants and animals which we know to have existed at different epochs. We shall thus learn, at least, some of the prevailing characteristics of the earth during its transitions, and be in a better condition for applying our knowledge of physical power to the explanation of the various geological phenomena.

Among the earliest races, we have those remarkable forms, the trilobites, inhabiting the ancient ocean.

These crustacea bear some resemblance, although a very remote one, to the common wood-louse, and, like that animal, they had the power of rolling themselves into a ball when attacked by an enemy. The eye of the trilobite is a most remarkable organ; and in that of one species, *Phacops caudatus*, not less than two hundred and fifty lenses have been discovered. This remarkable optical instrument indicates that these creatures lived under similar conditions to those which surround the crustacea of the present day.

At the period of the trilobites of the Silurian rocks, all the animals contemporaneous with them had the organs necessary for the preservation of life in the waters.

Next in order of time to the trilobite, the most singular animals inhabiting those ancient seas, whose remains have been preserved, are the *Cephalopoda*, possessing some traces of organs which belong to vertebrated animals. There are numerous arms for locomotion and prehension, arranged in a centre round the head, which is furnished with a pair of sharp, horny mandibles, embedded in powerful muscles. These prehensile arms are provided with a double row of suckers, by which the animal seized its prey. Of these cephalopodous animals there are many varieties, but all of them appear to be furnished with powers of rapid locomotion, and those with shells had an hydraulic arrangement for sinking themselves to any depth of the seas in which, without doubt, they reigned the tyrants.

Passing by without notice the numerous fishes which appear to have exhibited a similar order of progression to the other animals, we must proceed to the more remarkable period when the dry land first began to appear.

All the animals found in the strata we have mentioned, are such as would inhabit the seas; but we gradually arrive at distinct evidence of the separation of the land from the water, and the "green tree, yielding seed," presents itself to our attention; not that the strata earlier than this are entirely destitute of any remains indicating vegetable growth, but those they exhibit are such as, in all probability, may be referred to marine plants.

Those plants, however, which are found in the carboniferous series are most of them distinguished by all the characteristics of those which grow upon the land; we, therefore, in the mutilated remains of vegetation left us in our coal-formations, read the history of our early world.

Then the reed-like calamite bowed its hollow and fragile stems over the edges of the lakes, the tree-ferns grew luxuriantly in the shelter of the hills, and gave a wild beauty to the humid valleys. The lepidodendrons spread themselves in mighty forests along the plains, which they covered with their curious cones; whilst the sigillariæ extended their multitudinous branches, wreathing like serpents amongst the luxurious vegetation, and embraced, with their roots (stigmariæ), a most extensive space on every side.([228])

The seas and lakes of this period abounded with minute animals nearly allied to the coral animals, which are now so actively engaged in the formation of islands in the tropical and southern seas. During the ages which passed by without any remarkable disturbance of the surface of the earth, the many bands of mountain limestone were formed by the ceaseless activity of these minute architects. Encrinites (creatures in some respects resembling star-fish) existed in vast numbers in the oceans of this time; and the great variety of bivalve shells, and those of a spiral character, discovered in the rocks of this period, show the waters of the newer palæozoic period to have been instinct with life.

In the world, then, as it does now, water acting on the dry land produced remarkable changes. We have evidence of extensive districts over which the most luxuriant vegetation must have spread for ages, — from the remains of plants in every state of decay, — which we find went to form our great coal-fields. These, by some changes in the relative levels of land and water, became covered with this fluid; and over this mass of decaying organic matter, sand and mud were for ages being deposited. At length, rising above the surface, it becomes covered with vegetation, which is, after a period, submerged; the same deposition of sand and mud again takes place, it is once more fitted for vegetable growth, and thus, cycle after cycle, we see the dry land and the water changing places with each other. This will be evident to every one who will carefully contemplate a section of one of the coal-fields of Great Britain. We find a stratum of coal lying upon a bed of underclay, and above it an extensive stratum of shale or sandstone, probably formed by the denudation of the neighboring hills; and in this manner we have many strata of coal, shale, clay, ironstone, and sandstone alternating with each other.

Ascending in the series, we have now formations of a more

recent character, in which fishes of a higher order of organization, creeping and flying saurians, crocodiles and lizards, tortoises, serpents, and frogs, are found. The lias formations (a term corrupted from *layers)*, consisting of strata in which an argillaceous character prevails, stand next in series. In these we have animals preserved in a fossil state, of a distinguishingly different character from those of the inferior strata. Corals are not found in the British Isles; but we meet with extended beds of pentacrinites, some inches in thickness; and their remains are often so very complete that every part of the skeleton can be made out, although so complicated that it cannot consist of less than 150,000 parts. In these formations we often find the curiously beautiful remains of the ammonites, of which an immense variety have been found. Of the belemnites — animals furnished with the shell and the ink-bag of the cuttle-fish, with which it darkened the water to hide itself from enemies — numerous varieties have also been discovered. In addition to these we find nautili; and sixty species of extinct fishes have been described by Agassiz from the lias of Lyme Regis alone.

When these rocks were in the progress of formation, there existed the ichthyosaurus, or fish-lizard, which appears, in many respects, to have resembled the crocodile of the Nile. It was a predatory creature of enormous power, and must have been the tyrant and terror of the seacoasts which it inhabited. Its alligator-like jaws, its powerful eye, its fish-like fins, and turtle-like paddles, were all formed to facilitate its progress as a destructive agent. The plesiosaurus was, if possible, a still more remarkable creation. To the head of a lizard was united an enormously long neck, a small and fish-like body, and the tail of a crocodile; it appears formed for existence in shallow waters, so that, when moving at the bottom, it could lift its head above the surface for air, or in search of its food. The flora of this period must have been tolerably extensive: and it resembled the vegetation which exists at present in tropical regions.

Races of reptiles still have place upon the earth, and we have now the megalosaurain remains, indicating a strength and rapacity which would render them objects of terror as well as astonishment, could they be restored to the world which they once ravaged. An enormous bat-like creature also existed at this time — the pterodactyl — which, in the language of Cuvier, was, "undoubtedly, the most extraordinary of all the beings of whose former existence a knowledge is granted to us, and that which, if seen alive, would appear most unlike anything that exists in the present world." "You see before you," says the same writer, "an animal which, in all points of bony structure, from the teeth to the extremities of the nails, presents the well-known saurian characteristics, and of which no one can doubt that its integuments and soft parts, its scaly armour and its organs of circulation and reproduction, are likewise analogous. But it was, at the same time, an animal provided with the means of flying; and, when stationary, its wings were probably folded back like those of a bird, although, perhaps, by the claws attached to its fingers, it might suspend itself from the branches of trees."([229])

From the disintegration of the older rocks have no doubt arisen those formations which are known as the oolitic series. In these strata are preserved the remains of plants and animals more resembling those which now exist upon the earth; and, for the first time, — unless the evidence of the footsteps of birds on the new red sandstone of America be accepted, — we meet with the remains of winged creations.

In these formations we discover animals belonging to the class mammalia, — the amphitherium and the phascolotherium, — which appear to have resembled, in many respects, the marsupial animals of New Holland.([230])

The wealden formations, which are the next in order of position, are a series of clays and sands, with subordinate beds of limestone, grit, and shale. These have, in some instances, been formed in the sea; but they may be usually regarded as fresh-water deposits. All the older rocks bear evident marks of marine origin, unless some of the coal-measure strata may be regarded as otherwise; but nearly all the wealden series contain the remains of land, fresh water, and estuary animals, and of land vegetables. The animals which we discover, preserved, to tell the history of this period, are numerous, and have marked peculiarities to distinguish them from those already described, or from any now existing on the earth. We find land saurians of a large kind, and animals of all sizes, even insects, of which a great variety are found in the wealds. The remarkable iguanodon was an animal which, even by the cautious measurement of Professor Owen, must have been at least twenty-eight feet long; and this enormous creature was suspected, by Cuvier, and has been proved by Owen, to have been "an herbivorous saurian for terrestrial life."([231]) Dr. Mantell calculates that no less than seventy individuals of the iguanodon of all ages have come under his notice; and the bones of a vast number of others must have been broken up by the workmen in the few quarries of Tilgate grit; so that these creatures were by no means rare at the period of their existence.([232])

The uppermost of these secondary formations is the cretaceous or chalk group, which spreads over a large portion of south-eastern England, and is met with in all parts of Europe. This chalk, which is a carbonate of lime, appears to have been slowly precipitated from tranquil water, as, according to Sir Henry De la Beche, organic remains are beautifully preserved in it. Substances of no greater solidity than common sponges retain their forms, delicate shells remain unbroken, fish even are frequently not flattened, and altogether we have the appearances which justify us in concluding that, since these organic exuviæ were entombed, they have been protected from pressure by the consolidation of the rock around them.([233])

Beneath the chalk exists what has been called, from its colour — derived from a silicate of the protoxide of iron, — green sand, and was, no doubt, formed by deposition from the same water in which the carbonate of lime was suspended, — the green sand falling to the bottom more readily from its greater specific gravity. "The tranquillity," observes Sir Henry De la Beche, "which seems to have prevailed during this great accumulation of siliceo-calcareous matter, whether it may have been a deposit from water, in which it was mechanically suspended, partly the work of living creatures, or in a great measure chemical, is very remarkable."([234])

In the chalk, the remains of the leaves of dicotyledonous plants and fragments of wood are found more abundantly than in the earlier strata, many of which are marked with the perforations of marine worms, indicating that they had floated for some time in the ocean. It should, however, be remembered, that leaves have been found in the new red sandstone; and the flora of the coal-formation must not be forgotten. The manner in which silica has deposited itself on organic bodies — the sponges — is curious; the whole of the organized tissue being often removed, and flint having taken its place. These flints abound in the upper chalk. The association of carbon and silicon, combined with oxygen, as we find them in the cretaceous formations, is most interesting, and naturally gives rise to some speculation on the relation of these two elements. Both carbon and silicon, as has been already shown, exist in several allotropic conditions; and, although the statements made by Dr. Brown relative to the conversion of carbon into silicon are proved to be grounded on experimental error, it is not improbable that a very intimate relation may exist between these elements.(235) The probability is, that the sponge animal has the power of secreting silica to give strength to its form. "Many species," says Rymer Jones, speaking of recent sponges, "exhibiting the same porous structure, have none of the elasticity of the officinal sponge — a circumstance which is due to the difference observable in the composition of their skeletons or ramified framework. In such, the living crust forms within its substance not only tenacious bands of animal matter, but great quantities of crystallized spicula, sometimes of a calcareous, at others of a siliceous, nature." Thus, a frame of siliceous matter being formed by the living animal, a deposition of the same substance is continued after death.

Sea-urchins, and star-fish, and numerous fossil shells, are found in these beds, which, however, differ materially from the remains of the same animals found in the earlier formations. A vast number of new species and genera of fish are also discovered in the chalk.

Nearly all the animals and plants which existed up to this period are now extinct, although they have some imperfect representatives at the present day.

The uppermost group, which has been called the supercretaceous or tertiary formations, appears in our island to have been formed during four great eras, as we find fresh-water deposits alternating with marine ones. The terms *eocene*, which is the first or oldest deposit; *miocene*, which is the second; *pliocene*, which is the third; and the *newer pliocene*, which is the fourth and last, have been applied to these formations, the names referring to the respective proportions of existing species found among their fossil shells.(236)

All these formations show distinct evidence of their having

been deposited from still or slowly-flowing deep waters. Thus the eocene appears in the Paris basin, formed clearly at an estuary, in which are mingled some interesting fresh-water deposits; in the lacustrine formations in Auvergne; also at Aix; and in the north of Italy. It appears probable that, in the formations generally termed eocene, both fresh-water and marine deposits have been confounded, and several formations of widely different eras regarded as the result of one. We have not yet been furnished with any distinct and clear evidence to show that the deposits of the Paris basin, and those of Auvergne, are of the same age. At all events, it is sufficient for our present purpose to know that they are the result of actions which are now as general as they were when the plastic clay of Paris, and its sulphate of lime, or the London clay, were slowly deposited.

As a general conclusion, we may decide that, at the eocene period, existing continents were the sites of vast lakes, rivers, and estuaries, and were inhabited by quadrupeds, which lived upon their thickly-wooded margins. Many remains, allied to those of the hippopotamus, have been found in the subsidences of this period.

Examples of the miocene or middle tertiary era are to be found in Western France, over the whole of the great valley of Switzerland, and the valley of the Danube. In these deposits we find the bones of the rhinoceros, elephant, hippopotamus, and the dinotherium, an extinct animal, possessing many very distinguishing features.([237])

The pliocene period has been termed the age of elephants, and is most remarkable for the great mastodons and gigantic elks, with other animals not very unlike those which are contemporaneous with man.

In the superficial structure of the earth, the diluvium, alluvium, peat and vegetable soil, we have a continuation of the history of the mountains of the earth and of its inhabitants, which has been so briefly sketched. They bring us up to the period when man appeared in the world, since whose creation it is evident no very extensive change has been produced upon the surface. We have viewed the phenomena of each great epoch, marked as they are by new creations of organized beings, and it would appear as if, through the whole series, from the primary rocks up to the modern alluvial deposits, a progressive improvement of the earth's surface had been effected, to fit it at last for the abode of the human race.

Thus have we preserved for us, in a natural manner, evidences which, if we read them aright, must convince us that the laws by which creation has ever been regulated are as constant and unvarying as the Eternal mind by which they were decreed. Our earth, we find, by the records preserved in the foundation-stones of her mountains, has existed through countless ages, and through them all exhibited the same active energies that prevail at the present moment. By precisely similar influences to those now in operation, have rocks been formed, which, under like agencies, have been covered with vegetation, and sported over by, to us, strange varieties of animal life. Every plant that has grown upon the earliest rocks which presented their faces to the life-giving sun, has had its influence on the subsequent changes of our planet. Each trilobite, each saurian, and every one of the mammalia which exist in the fossil state, have been small laboratories in which the great work of eternal change has been carried forward, and, under the compulsion of the strong laws of creation, they have been made ministers to the great end of forming a world which might be fitting for the presence of a creature endued with a spark taken from the celestial flame of intellectual life.

For a few moments we will return to a consideration of the

operations at present exhibiting their phenomena, and examine what bearing they have upon our knowledge of geological formations.

During periods of immense, but unknown, duration, the ocean and the dry land are seen to have changed their places. Enormous deposits, formed at the bottom of the sea, are lifted by some mechanical, probably volcanic, force above the waters, and the land, like the ocean surrounding it, teems with life. This state of things lasts for ages; but the time arrives when the ocean again floods the land, and a new state of things, over a particular district, has a beginning.

It must not be imagined that the changes which we have spoken of, as if they were the result of slow decay and gradual deposit, were effected without occasional violent convulsions. Many of the strata which were evidently deposited at the bottom of the sea, and, of course, as horizontal beds, are now found nearly vertical. We have evidence of strata of immense thickness having been subjected to forces that have twisted and contorted them in a most remarkable manner. Masses of solid rock, many thousand feet deep, are frequently bent and fractured throughout their whole extent. Mountains have been upheaved by internal force, and immense districts have suddenly sunk far below their usual level. By the expansive force of that temperature which must be required to melt basaltic and trap rocks, the whole of the superficial crust of a country has been heaved to a great height, immense fissures have been formed by the breaking of the mass, and the melted matter has been forced through the opening, and overflowed extensive districts, or volcanoes have been formed, and wide areas have been buried under the ashes ejected from them. With the cause of these convulsions we are, at present, unacquainted.

We have evidence of the extent to which these forces may be

exerted, in the catastrophes which have occurred within historical times, and which have happened even in our own day. Herculaneum and Pompeii, buried under the lava and ashes of Vesuvius, in an hour when the inhabitants of these cities were unprepared for such a fearful visitation, — the frightful earthquakes which have, from time to time, occurred in South America, — are evidences of the existence of hidden forces which shake the firm-set earth. Similar ravaging catastrophes may have often occurred, and, involving cataclysms, swept the surface to produce the changes we detect over every part of the earth, compared with which the earthquakes and floods of history are but trivial things. Evidence has been adduced, to show that the mountains of the Old World may have approached in height the highest of the Andes or Himalayas, and these have not been destroyed by any sudden effect, but by the slow disintegrating action of the elements.([238]) All these phenomena are now in progress: the winds and the rains wear the faces of the exposed rock; their *débris*, mixed with decayed vegetable and animal matter, are washed off from the surface, and borne away by the rivers, to be deposited by the seas; — thus it is that the great delta of the Ganges is formed, and that a continual increase of matter is going on along the sides and at the mouths of rivers. The Amazon, the Mississippi, and other great rivers, bear into the ocean, daily, thousands of tons of matter from the surface of the earth.([239]) This is, of course, deposited at the bottom of the sea, and it must, in the process of time, alter the relative levels of the ocean and the land. Islands have been lifted by volcanic power from the bottom of the sea, and many districts in South America have been depressed by the same causes.

Changes as extensive have been, in all probability, effected by forces "equally or more powerful, but acting with less irregularity, and so distributed over time as to produce none of those interregnums of chaotic anarchy which we are apt to think (perhaps erroneously) great disfigurements of an order so beautiful and harmonious as that of nature."([240]) These forces are, without doubt, even now in action.

Had it not been for these convulsive disturbances of the surface, the earth would have presented an almost uniform plain, and it would have been ill adapted for the abode of man. The hills raised by the disturbances of nature, and the valleys worn by the storms of ages, minister especially to his wants, and afford him the means of enjoyment which he could not possess had the surface been otherwise formed. The "iced mountain-tops," condensing the clouds which pass over them, send down healthful streams to the valleys, and supply the springs of the earth, thus securing the fertility and salubrity of lands far distant. The severities of climate are mitigated by these conditions, and both the people of the tropics and those dwelling near the poles are equally benefited by them.

Gravitation, cohesion, chemical force, heat, and electricity must, from that hypothetical time when the earth floated a cloud of nebulous vapour, in a state of gradual condensation up to the present moment, have been exercising their powers, and regulating the mutations of matter.

When the dry land was beneath the waters, and when darkness was upon the face of the deep, the same great operations as those which are now in progress in the depths of the Atlantic, or in the still waters of our inland lakes, were in full activity. At length the dry land appears; and — mystery of mysteries — it soon becomes redolent of life in all the forms of vegetable and animal beauty, under the aspect of the beams of a glorious sun.

Geology teaches us to regard our position upon the earth as one far in advance of all former creations. It bids us look back through the enormous vista of time, and see, shining still in the remotest distance, the light which exposes to our vision many of nature's holy wonders. The elements which now make up this strangely beautiful fabric of muscle, nerves, and bone, have passed through many ordeals, ere yet it became fashioned to hold the human soul. No grain of matter has been added to the planet, since it was weighed in a balance, and poised with other worlds. No grain of matter can be removed from it.

Under the forces we have been considering, acting as so many contending armies, matter passes from one condition to another, and what is now a living and a breathing creature, or a delicate and sweetly-scented flower, has been a portion of the amorphous mass which once lay in the darkness of the deep ocean, and it will again, in the progress of time, pass into that condition where no evidences of organization can be found, — again, perhaps, to arise clothed with more exalted powers than even man enjoys.

When man places himself in contrast with the Intelligences beyond him, he feels his weakness; and the extent of power which he can discover at work, guided by a mysterious law, is such, that he is dwarfed by its immensity. But looking on the past, surveying the progress of matter through the inorganic forms up to the higher organizations, until at length man stands revealed as the chief figure in the foreground of the picture, the monarch of a world on which such elaborate care has been bestowed, and the absolute ruler of all things around him, he rises like a giant in the conscious strength of his far-searching mind. That so great, so noble a being should suffer himself to be degraded by the sensualities of life to a level with the creeping things, upon which he has the power to tread, is a lamentable spectacle, over which angels must weep.

The curious connection between the superstitions of races, the traditionary tales of remote tribes, and the developments of the truths of science, are often of a very marked character, and they cannot but be regarded as instructive. In the wonders of "olden time" fiction has ever delighted; and a thousand pictures have been produced of a period when beings lived and breathed upon the earth which have no existence now.

Hydras, harpies, and sea-monsters figure in the myths of antiquity. In the mythology of the northern races of Europe we have fiery-flying dragons, and poetry has placed these as the guardians of the "hoarded spirit," and protectors of the enchanted gold.

Through the whole of the romance period of European literature, nothing figures but serpents, "white and red," toiling and fighting under ground, — thus producing earthquakes, as in the story of Merlin and the building of Stonehenge; — and flying monsters, griffins and others, which now live only in the meaningless embellishments of heraldry. Curious it is, too, to find the same class of ideas prevailing in the East; — the monster dragons of the Chinese, blazoned on their standards and ornamenting their temples; — the Buddaical superstition that the world is supported on a vast elephant, which stands on the back of a tortoise, which again rests on a serpent, whose movements produce earthquakes and violent convulsions; — the rude decorations also of the temples of the Aztecs, which have been so recently restored to our knowledge by the adventurous travellers of Central America, — all give expression to the same mythological idea.

Do not all these indicate a faint and shadowy knowledge of a previous state of organic existence? The process of communion between man of the present, and the creations of a former world, we know not; it is mysterious, and forever lost to us. But even the most ignorant and uncultivated races of mankind have figured for themselves the images of creatures which, whilst they do really bear some resemblance to things which have for-ever passed away, do not, in the remotest degree, partake of any of the peculiarities of existing creations.

The ichthyosaurus, and the plesiosaurus, and the pterodactylus, are preserved in the rude images of harpies, of dragons, and of griffins; and, although the idea of an elephant standing on the back of a tortoise was often laughed at as an absurdity, Captain Cautley and Dr. Falconer at length discovered in the hills of Asia the remains of a tortoise in a fossil state of such a size that an elephant could easily have performed the feat. (241)

Of the ammonites, we have more exact evidence; they were observed by our forefathers, and called by them snake-stones; and according to the legends of Catholic saints they were considered by them as possessing a sacred character:—

> "Of these and snakes, each one
> Was changed into a coil of stone
> When holy Hilda prayed."

And in addition to this petrifying process, one of decapitation is said to have been effected; hence the reason why these snake-stones have no heads.

We also find, in the northern districts of our island, that the name of "St. Cuthbert's beads" is applied to the fossil remains of encrinites.

Thus we learn that to a great extent fiction is dependent upon truth for its creations; and we see that when we come to investigate any wide-spread popular superstition, although much distorted by the medium of error through which it has passed, it is frequently founded upon some fragmentary truth. There are floating in the minds of men certain ideas which are not the result of any associations drawn from things around; we reckon them amongst the mysteries of our being. May they not be the truths of a former world, of which we receive the dim outshadowing in the present, like the faint lights of a distant Pharos, seen through the mists of the wide ocean?

Man treads upon the wreck of antiquity. In times which are so long past, that the years between them cannot be numbered by the aids of our science, geology teaches us that forms of life existed perfectly fitted for the conditions of the period. These performed their offices in the great work; they passed away, and others succeeded to carry on the process of building a world for man. The past preaches to the present, and from its marvellous discourses we venture to infer something of the yet unveiled future. The forces which have worked still labour: the phenomena which they have produced will be repeated.

Ages on ages slowly pass away,
And nature marks their progress by decay.
The plant which decks the mountain with its bloom,
Finds in the earth, ere long, a wasting tomb:
And man — the great, the generous, and the brave —
Seeks in the soil, at last, a silent grave.
The chosen labours of his teeming mind
Fade with the light, and crumble with the wind;
And e'en the hills, whose tops appear to shroud
Their granite peaks deep in the vapoury cloud,
Melt slowly down to fill th' extended plains,
Worn by the breezes — wasted by the rains.
Earth sinks in ocean — seas o'erwhelm the land;
But 'neath the powers of the empyreal band —
Who, ever working at creation's wheel,
From the rude wrecks of matter still reveal
Forms of excelling beauty — earth will rise
Pure as the flame from love's own sacrifice,
And, beaming with the brightest smile of youth,
Proclaim mutation as the eternal truth.

CHAPTER XIV.

PHENOMENA OF VEGETABLE LIFE.

Psychology of Flowers — Progress of Matter towards Organization — Vital Force — Spontaneous Generation — The Vegetable Cell — Simplest Development of Organization — The Crystal and the Cell — Primitive Germ — Progress of Vegetation — Influence of Light — Morphology — Germination — Production of Woody Fibre — Leaves — Chlorophylle — Decomposition of Carbonic Acid — Influence of Light, Heat, and Actinism on the Phenomena of Vegetable Life — Flowers and Fruits — Etiolation — Changes in the Sun's Rays with the Seasons — Distribution of Plants — Electrical and Combined Physical Powers.

The variety of beautiful forms which cover the surface of this sphere serve, beyond the physical purposes to which we have already alluded, to influence the mind, and give character to the inhabitants of every locality. There are men who appear to be dead to the mild influences of flowers; but these sweet blossoms — the stars of our earth — exert a power as diffusive as their pervading odours.

The poet tells us of a man to whom

> The primrose on the river's brim
> A yellow primrose was to him,
> And it was nothing more.

But it was something more. He, perhaps, attended not to the eloquent teaching of its pure, pale leaves: he might not have been conscious of the mysterious singing of that lowly flower: he might, perchance, have crushed it beneath his rude foot rather than quaff the draught of wisdom which it secreted in its cell; but the flower still ministered to that mere sensualist, and in its strange, tongueless manner, reproved his passions, and kept him "a wiser and a better man," than if it had pleased God to have left the world without the lovely primrose.

The psychology of flowers has found many students — than whom not one read them more deeply than that mild spirit who sang of the Sensitive Plant, and in wondrous music foreshadowed his own melancholy fate.(242) That martyr to sensibility, Keats, who longed to feel the flowers growing above him, drew the strong inspiration of his volant muse from those delicate creations which exhibit the passage of inorganic matter into life; and other poets will have their sensibilities awakened by the æsthetics of flowers, and find a mirror of truth in the crystal dew-drop which clings so lovingly to the purple violet.

If we examine carefully all the conditions of matter which we have made the subject of our studies, we cannot but perceive how gradual is the progress of the involved action of the physical forces, as we advance from the molecule — the mere particle of matter — up to the organic combination. At first, we detect only the action of cohesion in forming the rude mass; then we have the influence of the crystallogenic powers giving a remarkable regularity to bodies; we next discover the influences of heat and electrical force in determining condition, and of chemical action as controlled by them. Yet, still, we have a body without organization. Light exerts its mysterious powers; and the same elements assume an organized form; and, in addition to the recognized agencies, we perceive others on which vitality evidently depends. These empyreal influences become more and more complicated to us: ascending in the scale, they rise beyond our science; and, at length, we find them guiding the powers of intelligence, while instinct and reason are exhibited in immediate dependence upon them.

Let it not be imagined that this view has any tendency to

materialism. The vital energy is regarded as a spiritualization, and reason as a divine emanation; but they are connected with materialities, on which they act, and by which they are themselves controlled. The organic combinations, and the physical powers by which these unions of matter are effected and retained, have a direct action over that ethereality which gives life, and the powers of life again control these more material forces. The spirit, in whatever state, when connected with matter, is like Prometheus chained to his rock, in a constant struggle to escape from his shackles, and assert the full power of its divine strength.

We have seen variety enough in the substances which make up the inorganic part of creation; but infinitely more varied are the forms of organization. In the vegetable world which is immediately around us, from the green slime of our marshes to the lustrous flowers of our gardens and the lordly trees of our forests, what an extraordinary diversity of form is apparent! From the infusoria of an hour, to the gigantic elephant roaming in his greatness in the forests of Siam — the noble lion of the caves of Senegal — the mighty condor of the Andes — and onward to man, the monarch of them all, how vast are the differences, and yet how complete are they in their conditions! In the creation we have examined, we have had conclusive evidence, that from the combination of atoms every peculiar form has been produced. In the creation we are about to examine, we shall discover that all the immense diversity of form, of colour, and condition which is spread over the world in the vegetable and animal kingdoms, results from the combination of cells. The atom of inorganic nature becomes a cell in organic creation. This cell must be regarded as the compound radical of the chemist, and by decomposing it, we destroy the essential element of organization.

With the mysterious process by which the atom is converted

into a cell, or a compound radical, we are unacquainted; but we must regard the cell as the organic atom. It is in vain that the chemist or the physiologist attempts to examine this change of the inorganic elements to an organized state; it is one of the mysteries of creation, which is to be, in all probability, hid from our eyes, until this "mortal coil" is shaken off, and we enjoy the full powers of intelligence in our immortal state.

Again and again has the attention of men been attracted to the *generatio æquivoca;* they have sometimes thought they have discovered a *generatio primitiva* or *spontanea;* but a more careful examination of these organisms has shown that an embryo existed — a real germination has taken place.

Count Rumford(243) stated that threads of silk and wool had the power of decomposing carbonic acid in water in the sunshine; and hence, some have referred organization to a mere chemical change produced by luminous excitation; and we have heard of animal life resulting from pounded siliceous matter. All such statements must be regarded as evidences of imperfect investigation.

Dr. Carus, alluding to the experiments of Gruithuisen, Priestley, and Ingenhousz,(244) says: "These show, more than any other experiments, that, in the purest water, under the influence of air, light, and heat, beings are formed, which, oscillating as it were between the animal and the plant, exhibit the primitive germs of both kingdoms."(245) Treviranus(246) repeated, and appeared to confirm these results; but in these experiments we have no evidence that the germ did not previously exist in the spring-water which was employed.

Some have regarded the cell as a crystal; they see the crystal forming, by the accumulation of atoms, into a fixed form, under the influence of an "inner life;" and, advancing but a step, they regard the cell as the result of an increased exercise of the physical influences.(247) We have referred crystalline form to certain magnetic conditions; and it is evident that every atom of the cell is influenced by similar conditions; but if we place a crystal in its natural fluid, though it increases in size, it never alters in form: whereas, if we place a cell in its natural position, it gives indications of motion, it unites with other cells, and we have a development of organs which are in no respect the same in form as the original. From a vesicle floating invisible to the unaided human sense in its womb of fluid, is produced a plant possessing strange powers, or an animal gifted with volition. The idea, that two kinds of polarity — light on one side, and gravitation on the other — produce the two peculiar developments of roots and branches, can only be regarded as one of those fanciful analogies which prove more imagination than philosophy.(248)

The conditions are, however, most curious; they deserve very attentive study; but in examining the phenomena, the safest course is to allow the effects as they arise to interpret to us, and not admit the love of hypothesis to lead us into bewildering analogies; or uncertain phenomena to betray us to hasty inferences. It is of this evil that Bacon speaks, in his "Advancement of Learning." He says: —

"The root of this error, as of all others, is this, that men, in their contemplations of nature, are accustomed to make too timely a departure, and too remote a recess from experience and particulars, and have yielded and resigned themselves over to the fumes of their own fancies and popular argumentations."

Without venturing, therefore, to speculate on the origin of the primitive cell, or unit of vegetable life, which involves the problem of the metamorphosis of a rude mass — the primitive transformation of the rudimentary atoms into organic form — we must admit that the highly organized plant is but an aggregation of these cells; their arrangement being dependent upon certain properties peculiar to them, and the exercise of forces such as we have been studying, — all of which appear to act externally to the plant itself.

Experiments have been brought forward in which it appeared that, after all organization which could by any possibility exist had been destroyed by the action of fire, solutions of flint and metallic salts have, under the influence of electric currents, exhibited signs of organic formations, and that, indeed, insects — a species of acari — have been developed in them. The experiments were made with care, and many precautions taken to cut off all chances of any error, but not all the precautions required in a matter of such exceeding delicacy; and we are bound not to receive the evidence yet afforded as the true expression of a fact without much further investigation. All experience, setting aside the experiment named, is against the supposition that pounded or dissolved flint could by any artificial means be awakened into life. Ova may have been conveyed into the vessels which contained the solutions under experiment; and in due time, although possibly quickened by electric excitation, the animals — the most common of insects — came into existence. ([249])

The rapid growth of confervæ upon water, has often been brought forward as evidence of a spontaneous generation, or the conversion of inorganic elements into organic forms; but it has been most satisfactorily proved that the germ must be present, otherwise no evidence of anything like organization will be deve oped. All the conditions required for the production of vegetable life appear to show, that it is quite impossible for any kind of plant, even the very lowest in the scale, to be formed in any other way than from an embryo in which are contained the elements necessary for it, and the arrangements required for the various processes which are connected with its vitality.

The earth is now covered with vegetable life, but there must have existed a time when "darkness was upon the face of the deep," and organization had not yet commenced tracing its lovely network of cells upon the bare surface of the ocean-buried rock. At length the mystery of organic creation began: into this science dares not penetrate, but it is privileged to begin its search a little beyond this point, and we are enabled to trace the progress of organic development through a chain of interesting results which are constantly recurring.

If we take some water, rising from a subterranean spring, and expose it to sunshine, we shall see, after a few days, a curious formation of bubbles, and the gradual accumulation of green matter. At first, we cannot detect any marks of organization — it appears a slimy cloud of an irregular and undetermined form. It slowly aggregates, and forms a sort of mat over the surface, which at the same time assumes a darker green colour. Careful examination will now show the original corpuscles involved in a network formed by slender threads, which are tubes of circulation, and may be traced from small points which we must regard as the compound atom, the vegetable unit. We must not forget, here, that we have to deal with four chemical elements, oxygen, hydrogen, carbon, and nitrogen, which compose the world of organized forms, and that the water affords us the two first as its constituents, gives us carbon in the form of carbonic acid dissolved in it, and that nitrogen is in the air surrounding it, and frequently mixed with it also.

Under the influence of the light, we have now seen these elements uniting into a mysterious bond, and the result is the formation of a cellular tissue, which possesses many of the functions of the noblest specimens of vegetable growth. But let us examine the progress. The bare surface of a rock rises above the waters covered over with this green slime, a mere veil of delicate network, which, drying off, leaves no perceptible trace behind it; but the basis of a mighty growth is there, and under solar influence, in the process of time, other changes occur.

After a period, if we examine the rock, we shall find upon its face little coloured cups or lines with small hard discs. These, at first sight, would not be taken for plants, but on close examination they will be found to be lichens. These minute vegetables shed their seed and die, and from their own remains a more numerous crop springs into life. After a few of these changes, a sufficient depth of soil is formed, upon which mosses begin to develope themselves, and give to the stone a second time a faint tint of green, a mere film still, but indicating the presence of a beautiful class of plants, which, under the microscope, exhibit in their leaves and flowers many points of singular elegance. These mosses, like the lichens, decaying, increase the film of soil, and others of a larger growth supply their places, and run themselves the same round of growth and decay. By and by, funguses of various kinds mingle their little globes and umbrella-like forms. Season after season plants perish and add to the soil, which is at the same time increased in depth by the disintegration of the rock over which it is laid, the cohesion of particles being broken up by the operations of vegetable life. The minute seeds of the ferns floating on the breeze, now find a sufficient depth of earth for germination, and their beautiful fronds, eventually, wave in loveliness to the passing winds.

Vegetable forms of a higher and a higher order gradually succeed each other, each series perishing in due season, and giving to the soil additional elements for the growth of plants of their own species or those of others. Flowering herbs find a genial home on the once bare rock; and the primrose pale, the purple foxglove, or the gaudy poppy, open their flowers to the joy of light. The shrub, with its hardy roots interlaced through the soil, and binding the very stones, rose rich in its bright greenery. Eventually, the tree springs from the soil, and where once the tempest beat on the bare cold rock, is now the lordly and branching monarch of the forest, with its thousand leaves, affording shelter from the storm for bird and beast.

Such are the conditions which prevail throughout nature in the progress of vegetable growth: the green matter gathering on a pond, the mildew accumulating on a shaded wall, being the commencement of a process which is to end in the development of the giant trees of the forest, and the beautifully tinted flower of nature's most chosen spot.

We must now consider closely the phenomena connected with the growth of an individual plant, which will illustrate the operation of physical influences throughout the vegetable world. The process by which the embryo, secured in the seed, is developed, is our first inquiry.

An apparently dead grain is placed in the soil. If the temperature is a few degrees above the freezing point, and holds a due quantity of water, the integument of the seed imbibes moisture and swells; the tissue is softened, and the first effort of vital force begins. The seed has now the power of decomposing water, the oxygen combines with some of the carbon of the seed, and is expelled as carbonic acid. This part of the process is but little removed from the merely chemical changes which we have already considered. We find the starch of the seed changed into sugar, which affords nutritive food for the developing embryo. The seed now lengthens downwards by the radicle, and upwards by the cotyledons, which, as they rise above the earth, acquire a green colour. Carbonic acid is no longer given off. These cotyledons, which are two opposite roundish leaves, act as the lungs; by them carbonic acid is conveyed to the roots, it is carried by a circulating process now in full activity through the young plant, it is deprived of its carbon, and oxygen is exhaled from it. The plant at this period is little more than an arrangement of cellular tissue, a very slight development of vascular and fibrous tissue appearing as a cylinder lying in the centre of the sheath. At this point, however, we begin distinctly to trace the operations of a new power; the impulses of life are evident.

The young root is now lengthening, and absorbing from the moisture in the soil, which always contains some soluble salts, a portion of its nutriment, which is impelled upwards by a force — probably capillary attraction and endosmose action combined — to the point from which the plumule springs. The plumule first ascends as a little twig, and, at the same time, by exerting a more energetic action on the carbonic acid than the cotyledons have done, the carbon retained by them being only so much as is necessary to form chlorophylle, or the green colouring matter of leaves, some wood is deposited in the centre of the radicle. From this time the process of lignification goes on through all the fabric, — the increase, and indeed the life, of the plant depending upon the development of a true leaf from the plumule.

It must not be imagined that the process consists, in the first place, of a mere oxidation of the carbon in the seed, — a slow combustion by which the spark of life is to be kindled; — the hydrogen of the water plays an important part, and, combining also with the carbon, forms necessary compounds, and by a secondary process gives rise again to water by combination with oxygen in the cells of the germinating grain. Nor must we regard the second class of phenomena as mere mechanical processes for decomposing carbonic acid, but the result of the combined influences of all the physical powers and life superadded.

This elongating little twig, the plumule, at length unfolds

itself, and the branch is metamorphosed into a leaf. The leaf aërates the sap it receives, effects the decomposition of the carbonic acid, the water, and in all probability the ammonia which it derives from the air, and thus returns to the pores, which communicate with the pneumatic arrangements of the plant, the necessary secretions for the formation of bark, wood, and the various proximate principles which it contains.

After the first formation of a leaf, others successively appear, all constructed alike, and performing similar functions. The leaf is the principal organ to the tree; and, indeed, Linnæus divined, and Goëthe demonstrated, the beautiful fact, that the tree was developed from this curiously-formed organ.

"Keeping in view," says the poet-philosopher, "the observations that have been made, there will be no difficulty in discovering the leaf in the seed-vessel, notwithstanding the variable structure of that part and its peculiar combinations. Thus the pod is a leaf which is folded up and grown together at its edges, and the capsules consist of several leaves grown together, and the compound fruit is composed of several leaves united round a common centre, their sides being opened so as to form a communication between them, and their edges adhering together. This is obvious from capsules which, when ripe, split asunder, at which time each portion is a separate pod. It is also shown by different species of one genus, in which modifications exist of the principle on which their fruit is formed; for instance, the capsule of *nigilla orientalis* consists of pods assembled round a centre, and partially united; in *nigilla damascena* their union is complete."([230])

Professor Lindley thus explains the same view: "Every flower, with its peduncle and bracteolæ, being the development of a flower-bud, and flower-buds being altogether analogous to leaf-buds, it follows as a corollary that every flower, with its peduncle and bracteolæ, is a metamorphosed branch.

"And, further, the flowers being abortive branches, whatever the laws are of the arrangement of branches with respect to each other, the same will be the laws of the flowers with respect to each other.

"In consequence of a flower and its peduncle being a branch in a particular state, the rudimentary or metamorphosed leaves which constitute bracteæ, floral envelopes, and sexes, are subject to exactly the same laws of arrangement as regularly-formed leaves."(251)

The idea that the leaf is the principal organ of the plant, and that from it all the other organs are probably developed, is worthy the genius of the great German poet.

Every leaf, a mystery in itself, is an individual gifted with peculiar powers; they congregate in families, and each one ministers to the formation of the branch on which it hangs, and to the main trunk of the tree of which it is a member. The tree represents a world, every part exhibiting a mutual dependence.

> "The one red leaf, the last of its clan,
> That dances as often as dance it can;
> Hanging so light and hanging so high,
> On the topmost twig that looks up at the sky,"

is influenced by, and influences, the lowest root which pierces the humid soil. Like whispering voices, the trembling leaves sing rejoicingly in the breeze and summer sunshine, and they tremble alike with agony when the autumnal gale rends them from the parent stalk. The influences which pervade the whole, making up the sum of vital force, are disturbed by every movement throughout the system; a wound on a leaf is known to disturb the whole, and an injury inflicted on the trunk interferes with the processes which are the functions of every individual leaf.(252)

The consideration of the physical circumstances necessary to germination and vegetable growth, brings us acquainted with many remarkable facts. At a temperature below the freezing point, seeds will not germinate; at the boiling point of water, a chemical change is produced in the grain, and its power of germinating is destroyed. Heat, therefore, is necessary to the development of the embryo, but its power must only be exerted within certain prescribed limits: these limits are only constant for the same class of seeds, they vary with almost every plant. This is apparent to every one, in the different periods required for germination by the seeds of dissimilar vegetables.

The seed is placed in the soil; shade is always — absolute darkness sometimes — necessary for the success of the germinating process. We have seen that the first operation of nature is peculiarly a chemical one, but this manifestation of affinity is due to an exertion of force, which is directly dependent upon solar power. If seeds are placed under all the necessary conditions of warmth and moisture, but exposed to unmixed light, they will not germinate; but if we obstruct the luminous rays, allowing the chemical power to act, which is to be done by the interposition of blue glass, the birth of the young plant proceeds without any interruption. But let us take a truly natural example. The seed is buried in the soil, when the genial showers of spring, and the increasing temperature of the earth, furnish the required conditions for this chemistry of life, and the plant eventually springs into sunshine. If, however, we place above the soil a yellow glass, — which we have shown possesses the property of separating light from actinism or chemical power, — and thus consequently insure the influence of only light and heat upon the soil, no seeds will germinate. If, on the contrary, a blue medium is employed, by which actinic power, freed from the interference of light, is rendered more active, germination takes place more readily than usual. Thus we obtain evidence that even through some depth of soil this peculiar solar power is efficient, and that under its excitement the first spring of life, in the germ, is effected.

The cotyledons and the plumule being formed, the plant undergoes a remarkable change. The seed, like an animal, absorbed oxygen and exhaled carbonic acid; the first leaves secrete carbon from carbonic acid inspired, and send forth, as useless to the plant, an excess of oxygen gas.

This power of decomposing carbonic acid is a vital function which belongs to the leaves and bark. It has been stated, on the authority of Liebig, that during the night the plant acts only as a mere bundle of fibres, — that it allows of the circulation of carbonic acid and its evaporation, unchanged. In his eagerness to support his chemical hypothesis of respiration, the able chemist neglected to inquire if this was absolutely correct. The healthy plant never ceases to decompose carbonic acid; but, during the night, when the excitement of light is removed, a much less quantity is decomposed, than when a stimulating sun, by the action of its rays, is compelling the exertion of every vital function.

During this process, we have another example of natural organic chemistry. The three inorganic elements of which the vegetable kingdom is composed, oxygen, hydrogen, and carbon, are absorbed as air or moisture by the leaves or through the roots, and the great phenomenon of vegetable life is the conversion of these to an organic condition. Sugar and gum are constantly produced, and from these, by combination with a little atmospheric nitrogen, a proteine compound is formed, which is an essential element in the progress of development. (253)

Plants growing in the light are beautifully green, the intensity of colouring increasing with the brilliancy of the light. Those which are grown in the dark are etiolated, their tissues are weak and succulent, their leaves of a pale yellow. It is, therefore, evident that the formation of this chlorophylle — as the green colouring matter of leaves is called — results from some action determined by the sun's rays.

Chlorophylle is a carbonaceous compound, formed in the leaves, serving, it would appear, many purposes in the process of assimilation. In the dark, the plant still requires carbon for its further development, and growing slowly, it removes it from the leaves, decomposing the chlorophylle, and supports its weak existence by preying on parts of its own structure, until at length, this being exhausted, it actually perishes of starvation.

This principle is effected in nature by the agency of light, *the luminous principle*, as distinguished from radiant heat, actinism, or electricity. That power which is most active in the development of the germ, will not produce the excitement necessary for the decomposition of carbonic acid and the secretion of carbon; and under the influence of radiations which have permeated blue media, the plant grows in a succulent state, the formation of wood being exceedingly small. Of course, each of the elementary forces plays an important part in the progress of growth: every power of the solar beam is necessary: the light to excite the plant to decompose carbonic acid, and heat and actinism to produce the formation of many of the peculiar juices natural to the various species. Plants always turn towards the light: the guiding power we know not, but the evidence of some impulsive or attracting force is strong; and the purpose for which they are constituted to obey it, is proved to be the dependence of vegetable existence upon luminous power.

Light is not, however, alone sufficient to perfect the plant: another agent is required to aid in the production of flowers and fruits, and this power is proved to be heat in some peculiar condition. Neither under the influence of the actinic or the luminous rays, as isolated by coloured media, will the plant produce flowers; but having reached that point of development when the reproductive functions are, by another change in the chemical operations going on within the vegetable structure, to be called forth, it has been found that the heat-rays, as completely separated as it is possible for them to be by red media, become in a remarkable manner effective. It has also been observed that plants bend from the red, or calorific rays, instead of towards them, as they are found to do to every other ray of the spectrum. From this we may argue that the influence of these rays is to check the vegetative processes, and thus to insure the perfection of the reproductive organs.

Observations, which have been extended over many years, prove that with the seasons these solar powers are, relatively to each other, subject to an interesting change. In the spring, the actinic power prevails, and during this period its agency is required for the development of the germ. As the summer comes on, the actinic rays diminish, and those of light increase. We see the necessity for this, since luminous power is required for the secretion of the carbon, with which the woody fibre is formed, and also for the elaboration of the proximate principles of the plant. Autumn, the season of fruit, is characterized by an increase of the heat-rays, and a diminution of the others: this change being necessary, as science now teaches us, for the due production of flower and fruit.

The calorific rays of the solar beam, to which the autumnal phenomena of vegetation appear particularly to belong, are of a peculiar character; they exhibit a curious compound nature, and, to distinguish them from the purely calorific principle, they have been called the Parathermic rays.([254]) To these rays we may refer the ripening of fruit and grain, and the browning of the leaf before its fall. May not the rise of the sap in spring be traced to the excitement of either light or actinism, and its recession, in the autumn, to that power from which the plant is found to bend, and which appears to be a modified form of heat?

There can be no doubt that the varieties of climate and the peculiarities of countries, as it regards their animal and vegetable productions, are dependent on the same causes. In every zone we find that vegetable organization is peculiarly fitted for the conditions by which it is surrounded. Under the equator, we have the spice-bearing trees, the nutmeg, the clove, the cinnamon, and the pepper-tree; there we have also the odoriferous sandal, the ebony, the banyan, and the teak; we have frankincense, and myrrh, and other incense-bearing plants; the coffee-tree, the tea-plant, and the tamarind.

A little further north we have the apricot, the citron, the peach, and the walnut. In Spain, Sicily, and Italy, we have the orange and lemon-tree blooming rich with perfume, and the pomegranate and the myrtle growing wild upon the rocks. Beyond the Alps, the vegetation again changes; instead of the cypress, the chestnut, and the cork-tree, which prevail to the south of them, we have the oak, the beech, and the elm. Still further north, we have the Scotch and spruce fir and larch. On the northern shores of the Baltic, and in that line of latitude, the hazel alone appears; and beyond this the hoary alder, the sycamore, and the mountain ash. Within the Arctic circle we find the mezereum, the water-lilies, and the globe-flowers; and, when the intensity of the northern cold becomes too severe even for these, the rein-deer moss still lends an aspect of gladness to the otherwise sterile soil.

The cultivation of vegetables depends on the temperature of the clime. The vine flourishes where the mean annual temperature ranges between 50° and 73°, and it is only cultivated profitably within 30° S. and 50° N. of the equator. To the same limits is confined the cultivation of maize and of olives. Cotton is grown profitably up to latitude 46° in the Old World, but only up to 40° in the New. We have evidence derived from photographic phenomena, that the constitution of the solar rays varies with the latitude. The effects of the sun's rays, in France and England, in producing chemical change, are infinitely more decided than, with far greater splendour of light, they are found to be in the lands under or near the equator.

Fungals are among the lowest forms of vegetation, but in these we have peculiarities which appear to link them with the animal kingdom. Marcet found that mushrooms absorbed oxygen, and disengaged carbonic acid. In all probability, this is only a chemical phenomenon of a precisely similar character to that which we know takes place with decaying wood. In the conversion of wood into *humus*, oxygen is absorbed, and combining with the carbon, it is evolved as carbonic acid. Of course, we have the peculiar condition of vitality to modify the effect, and we have, too, in this class of plants, the existence of a larger quantity of nitrogen than is found in any other vegetating substance.

These few sketches of remarkable phenomena connected with vegetation, are intended to show merely the operations of the physical powers of the universe, so far as we know them, upon these particular forms of organization. During the process of germination, electricity is, according to Pouillet, evolved; and again, in ripening fruits, there appears to be some evidence of electrical currents. Vegetables are, however, in the growing state, such good conductors of electricity, that it is not, according to the laws of this force, possible that they should accumulate it, so that the luminous phenomena stated to have been observed cannot be due to this agency. We know, however, that under every condition of change, whether induced by chemical or calorific action, electricity is set in motion; and we have reasons for believing that the excitation of light will also give rise to electrical circulation.

The question, whether plants possess sensation, whether they have any disposition of parts at all analogous to the nervous system of animals, has been often put forward, but as yet the answers have been unsatisfactory. The point is one well worthy all the attention of the vegetable physiologist; but, regarding plants as the link between the animal and the mineral kingdom, — looking upon phyto-chemistry, as exhibited by them, as the means employed to produce those more complex organizations which exist in animals, — we necessarily consider plants as mere natural machines for effecting organic arrangements, and, as such, that they cannot possess any nervous sensibility. Muscular contraction may be represented in many of their marvellous arrangements; and any disturbance produced by natural or artificial means, would consequently effect a change in the operation of those forces which combine to produce vegetable life. Indeed, the experiments of Carlo Matteucci, already referred to, prove that an incision across a leaf, the fracture of a branch, or the mere bruising of any part of the plant, interferes with the exercise of that power which, under the operation of luminous agency, decomposes carbonic acid, and effects the assimilation of the other elements.

To recapitulate. A plant is an organized creation; it is so in virtue of certain strange phyto-chemical operations, which are rendered active by the solar influences involved in the great phenomena of light, and by the excitation of caloric force and electrical circulation. It is a striking exemplification of the united action of certain empyreal powers, which give rise to the combination of inorganic principles under such forms that they become capable of obeying the mysterious impulses of life.

The poet has imaged the agency of external powers in various shapes of spiritualized beauty. From the goddess Flora, and her attendant nymphs, to the romantic enchantress who called up flowers by the light touch of her wand, we have, in all these creations, foreshadowings of the discovery of those powers which science has shown are essential to vegetable life. A power from without influences the plant; but the animal is dependent upon a higher agency which is potent within him.

The poet's dream pleases the imaginative mind; and, associating in our ideas all that is graceful and loveable in the female form, with that diviner feeling which impresses the soul with the sense of some unseen spirituality, we perceive in the goddess, the enchantress, or the sylph, pure idealizations of the physical powers. The spirit floating over these forms of beauty, and adorning them with all the richness of colour — painting the rose, and giving perfume to the violet — is, in the poet's mind, one which ascends to nearly the highest point of etherealization, and which becomes, indeed, to him, a spirit of light; they ride upon the zephyrs, and they float, in all the luxury of an empyreal enjoyment, down to the earth upon a sunbeam. Such is the work of the imagination. What is the result of the search of plodding science after truth? The sunbeam has been torn into rays, and every ray tasked to tell of its ministry.

Nature has answered to some of the interrogations; and, passing over all the earth, echoed from plant to plant, we have one universal cry, proclaiming that every function of vegetable life is due to the spirits of the sun.

The mighty Adansonia of Senegal, hoary with the mosses of five thousand years, — the Pohon upas in their deadly valleys, — the climbing lianas of the Guiana forests, — the contorted serpent-cactus on the burning hills, — the oaks, which spread their branches in our tempered climes, — the glorious flowers of the intertropical regions, and those which gem our virent plains, — the reindeer lichen of northern lands, and the confervæ of the silent pool, — the greatest and humblest creations of the vegetable world, — all proclaim their direct dependence upon the mysterious forces which are bound together in the silver thread of light.

CHAPTER XV.

PHENOMENA OF ANIMAL LIFE.

Distinction between the Kingdoms of Nature — Progress of Animal Life — Sponges — Polypes — Infusoria — Animalcula — Phosphorescent Animals — Annelidans — Myriapoda — Animal Metamorphoses — Fishes — Birds — Mammalia — Nervous System — Animal Electricity — Chemical Influences — Influence of Light on Animal Life — Animal Heat — Mechanical Action — Nervous Excitement — Man and the Animal Races, &c.

"A STONE grows; plants grow and live; animals grow, live, and feel." Such were the distinctions made by Linnæus, between the conditions of the three kingdoms of nature. We cannot, however, but regard them as somewhat illogical. The stone — a solid mass of unorganized particles — enlarges, if placed in suitable conditions, by the accretion of other similar particles around it; but it does not, according to any of the senses in which we use the word, *grow*. Plants and animals grow; and they differ, probably, only in the phenomena of sensation. Yet, the trembling mimosa, and several other plants, appear to possess as much feeling as sponges and some of the lower classes of animals. By this definition, however, of the celebrated Swedish naturalist, we have a popular and simple expression of a great fact.

As we have only to examine the question of the agency of the physical forces upon animal life, we must necessarily confine our attention to the more striking phenomena with which science has made us acquainted; and, having briefly traced the apparent order in which the advance of organization proceeded, we must direct our few concluding remarks to the physico-physiological influences, which we must confess to know but too imperfectly.

We learn that, during the states of progress which geology, looking into the arcana of time, has made us acquainted with, a great variety of animal forms were brought into existence. They lived their periods. The conditions of the surface, the sea, or the atmosphere were altered; and, no longer fitted for the enjoyments of the new life, these races passed away, and others occupied their places, which, in turn, went through all the stages of growth, maturity, and decay; until at length, the earth being fitted for the abode of the highest order of animals, they were called into existence; and man, the intellectual monarch of the world, was placed supreme amongst them all. Types of nearly all those forms of life which are found in the fossil state are now in existence; and if we examine the geographical distribution of animals — the zones of elevation over the surface of the earth, and the zones of depth in the ocean, — we shall find, now existing, animal creations strikingly analogous to the primitive forms and conditions of the earth's inhabitants. From the depths of the ocean we may even now study — as that most indefatigable naturalist, Professor Edward Forbes, has done — the varying states of organization under the circumstances of imperfect light and varying temperature.([235])

The gradual advance of animal life in the ascending strata has led to many speculations, ingenious and refined, on the progressive development of animals. That the changes of the inorganic world impressed new conditions on the organic structures of animals, to meet the necessities of their being, must be admitted. Comparative anatomy has demonstrated that such supposed differences really existed between the creatures of the secondary formations and those of the tertiary and the present periods. It has been imagined, but upon debatable foundations, that the atmosphere, during the secondary periods, was highly charged with carbonic acid; and, consequently, that though beneficial to the growth of plants, and peculiarly fitted for the conditions required by those which the fossil flora makes us acquainted with, it was not adapted to support any animals above the slow-breathing, cold-blooded fishes and reptiles. Under the action of the super-luxuriant vegetation of these periods, this carbonic acid is supposed to have been removed, and an addition of oxygen furnished; and thus, consequently, the earth gradually fitted for the abode of warm-blooded and quick-breathing creatures. We do, indeed, find a very marked line between the fossil remains of the lias formations which enclose the saurians, and the wealden, in which birds make their appearance more numerously than in any previous formation.

Founded upon these facts, speculations have been put forth on the gradual development of animals from the lowest up to the highest orders. Between the polype and man a continuous series has been imagined, every link of the chain being traced into connection with the one immediately succeeding it; and, through all the divisions, zoophytes, fishes, amphibia, reptiles, birds, and mammalia are seen, according to this hypothesis, to be derived by gradual advancement from the preceding orders. The first having given rise to amphibia, the amphibion gives birth to the reptile, the reptile advances to the bird, and from this class is developed the mammal. A slight investigation will convince us that this view has no foundation. Although a certain relationship may be found between some of the members of one class, and those of the other immediately joining it, yet this is equally discovered to exist towards classes more remote from each other; and in no one instance can we detect anything like the passage of an animal of one class into an animal of another; and until this is done, we cannot but regard the forms of animal life as distinct creations, each one fitted for its state of being, springing from the command of the great First Cause.(256)

But it is time we quit these speculative questions, and proceed to the examination of the general conditions of animal life at the present time.

Lowest in the scale of animals, and scarcely distinguishable from a vegetable, we find the sponge, attached to and passing its life upon a rock, exhibiting, indeed, less signs of feeling than many of the vegetable tribes. The chemical differences between vegetables and sponges are, however, very decided; and we find in their tissues a large quantity of nitrogen, a true animal element, which exists, but in smaller quantities, in vegetables.

These creations, standing between vegetable and animal life, possess the singular power of decomposing carbonic acid, as plants; and the water in which they live always contains an excess of free oxygen.

The polypes are a remarkably curious class. "Fixed in large arborescent masses to the rocks of tropical seas, or in our own climate attached to shells or other submarine substances, they throw out their ramifications in a thousand beautiful and plant-like forms; or, incrusting the rocks at the bottom of the ocean with calcareous earth, separated from the water which bathes them, they silently build up reefs and shoals, justly dreaded by the navigator; and sometimes giving origin, as they rise to the surface of the sea, to islands, which the lapse of ages clothes with luxuriant verdure, and peoples with appropriate inhabitants."(257) Most of the polypes are fixed and stationary; but the hydra and some others have the power of changing their positions, which they do in search of the light of the sun. They do not appear to have organs of sight requiring light; but still they delight in the solar influences. The most extraordinary fact connected with the hydra is its being multiplied by division. If an incision be made in the side of a hydra, a young polype soon developes itself; and if one of these creatures be divided, it quickly restores the lost portion of its structure. The varieties of the polypes are exceedingly numerous, and many of them are in the highest degree curious, and often very beautiful. The actiniæ, like flowers, appear to grow from the rocks, unfolding their tentacula to the light; and, in their eagerness for prey, they exhibit a beautiful play of colours, and most interesting forms. Microscopic zoophytes of the most curious shapes are found, — all of which attest, under examination, the perfection of all created things.

Infusoria and animalcula, — animals, many of them, appearing under the microscope as little more than a transparent jelly, — must be recognised as the most simple of the forms of life. They exist in all waters in uncountable myriads; and, minute creatures as they are, it has been demonstrated that many of the great limestone hills are composed entirely of their remains.

The acalephæ, or the phosphorescent animals of the ocean, are no less curious. From creatures of the most minute size, they extend to a considerable magnitude, yet they appear to be little more than animated masses of sea-water. If any one of these sea-jellies, or jelly-fishes, as they are often called, is cast upon the shore, it is soon, by the influence of the sun, converted into a mere fibre no thicker than a cobweb: an animal weighing seven or eight pounds, is very soon reduced to little more than as many grains. There are numerous varieties of these singular creatures, most of which are remarkable for the powerful phosphorescent light they emit. The beroes and the pulmonigrade shine with an intense white light many feet below the surface, whilst the *Cestum Veneris*, or girdle of Venus, gliding rapidly along, presents, on the edge of the wave, an undulating riband of flame of considerable length. There can be no doubt that this arises from the emission of phosphorescent matter of an unknown kind from the bodies of these animals.

The microscope has made us familiar with the mysteries of a minute creation which we should not otherwise have comprehended. These creatures are found inhabiting the waters and the land, and they exist in the intestinal structure of plants and animals, preying upon the nutritive juices which pass through their systems. Although these beings are so exceedingly small that even the most practised observer cannot detect them with the naked eye, they are proved, by careful examination under the microscope, to be in many cases elaborately organized. Ehrenberg has discovered in them filamentary nerves and nervous masses, and even vessels appropriated to the circulation of fluids, showing that they belong really to a high condition of existence.

Passing over many links in that curious chain which appears to bind the animal kingdom into a complete whole, we come to the articulata of Cuvier — the homogangliata of Owen.

All those creatures which we have been hitherto considering, are too imperfect in the construction of their simple organizations to maintain a terrestrial existence; they are, therefore, confined to a watery medium. In the articulata, we have evidences of higher attributes, and indications of instincts developed in proportion to the increased perfection of organization. Commencing with the annelidans, all of which, except the earthworms, are inhabitants of the waters, we proceed to the myriapoda, presenting a system intermediate in every respect between that of worms and insects; we then find embraced in the same order, the class insecta, which includes flies and beetles of all kinds; and, as the fourth division of articulated beings, the arachnidans, or spiders; and, lastly, the marine tribe of crustaceans.

The most remarkable phenomena connected with these animals are the metamorphoses which they undergo. The female butterfly, for instance, lays eggs, which, when hatched, produce caterpillars: these live in this state for some time, feeding upon vegetables, and, after casting their skins as they increase in size, at last assume an entirely different state, and, dormant in their oblong case, they appear like dead matter. This chrysalis, or pupa, is generally preserved from injury by being embedded in the earth, from which, after a season, a beautifully perfect insect escapes, and, floating on the breeze of summer, enjoys its sunshine, and revels amidst its flowers.

No less remarkable is the metamorphosis of the caducibranchiate amphibia, passing through the true fish condition of the tadpole to the perfect air-breathing and four-footed animal, the frog.

A metamorphosis of the crustaceans, somewhat similar to that which takes place in insects, has been of late years creating much discussion amongst naturalists: but the question appears to be now settled by the careful and long-continued observations of Mr. Thompson, and Mr. R. C. Couch.

A wide line of demarcation marks the separation of the invertebrata from the four great classes of vertebrate animals — fishes, reptiles, birds, and mammalia. Every part of the globe, — the ocean and the inland lake, — the wide and far-winding river, and the babbling stream, — the mountain and the valley, — the forest with its depth of shade, and the desert with its intensity of light, — the cold regions of the frost-chained north, and the fervid clime within the tropics — presents for our study innumerable animals, each fitted for the conditions to which it is destined; and through the whole we find a gradual elevation in the scale of intelligence, until at last, separated from all by peculiar powers, we arrive at man himself. In each of these four classes, the animals are furnished with a bony skeleton, which is in the young animal little more than cartilage; but, as growth increases, lime becomes deposited, and a sufficient degree of hardness is thus produced to support the adult formation. Some anatomists have endeavoured to show that even in the mechanical structure of the bony fabrics of animals, we are enabled to trace a gradual increase in the perfection of arrangement, from the fish until the most perfect is found in man. Many of the mammalia, however, are furnished with skeletons which really surpass that of man. These belong to animals which depend for subsistence upon their muscular powers, and with whom man is, in this particular, on no equality. What is the lord of the creation, compared with the antelope for fleetness, or with the elephant and many other animals for strength?

As we ascend the scale of animal life, we find a more perfectly developed nervous system; and the relative size of the brain, compared with that of the brute, is found progressively to increase, until it arrives at the utmost perfection in man. On the system of nerves depends sensation, and there can be no doubt that the more exalted the order of intelligence displayed, the more exquisitely delicate is the nervous system. Thus, in this world, refined genius must necessarily be attended with a condition of sensibility which, too frequently to the possessor, is a state of real disease.

It must be evident to every reader that but very few of the striking features of animal life have been mentioned in the rapid survey which has been taken of the progress of animal organization. The subject is so extensive that it would be quite impossible to embrace it within any reasonable limits; and it furnishes matter so curious and so instructive, that, having once entered on it, it would have been difficult to have made any selection, and we must have devoted a volume to the Æsthetics of natural science. Passing it by, therefore, with the mere outline which has been given, we must proceed to consider some of the conditions of vitality.

Bell has proved that one set of nerves is employed in conveying sensation to the brain, and another set in transferring the desires of the will to the muscles. By the separation of the main branch of one of the nerves of sensation, although all the operations of life will still proceed, the organ to which that nerve goes is dead to its particular sense. In like manner, if one of the nerves of volition is divided, the member will not obey the inclination of the brain. It is evident, therefore, although many of the great phenomena of vital force are dependent on the nervous system, and the paralysis of a member ensues upon the separation or the disease of a nerve, that the nerves are but the channels through which certain influences are carried. The *vis vitæ*, or vital principle — for we are compelled by the imperfection of our knowledge to associate under this one term the ultimate causes of many of the phenomena of life — is a power which, although constantly employed, has the capability of continually renewing itself by some inexplicable connection existing between it and many external influences. We know that certain conditions are necessary to the health of animals. Diseased digestion, or any interruption in the circulation of the blood, destroys the vital force, and death ensues. The processes of digestion and of the circulation are perfectly understood, yet we are no nearer the great secret of the living principle.

Animals are dependent on several external agents for the support of existence. The oxy en of the air is necessary for respiration. Animal heat, as will be shown presently, is in a great measure dependent upon it. The external heat is so regulated that animal existence is comfortably supported. Electricity is without doubt an essential element in the living process; and, indeed, many physiologists have been inclined to refer vital force to the development of electricity by chemical action in the brain; for which, however, there is no foundation in experiment.

The phenomena of the Torpedo and Gymnotus we have already noticed,([258]) and there are other creatures which certainly possess the power of secreting and discharging electricity. Galvani's experiments, and those of Aldini, appear to show—and the more delicate researches of Matteucci have satisfactorily determined — that currents of electricity are always circul.ting in the animal frame, that positive electricity is constantly passing from the interior to the exterior of a muscle ; and Matteucci, by arranging a series of muscles, has formed an electric pile of some energy.([259]) These currents have been detected in man, in pigeons, fowls, eels, and frogs.

In the human body, it is evident a large quantity of electricity exists in a state of equilibrium. This, however, may be said of every substance. It is perhaps more easily disturbed in the human system ; indeed, the manifestation of sparks from the hair and other parts of the body by friction is not uncommon. Every chemical action, it has been already shown, gives rise to electrical manifestations ; and the animal body is a laboratory, beautifully fitted with apparatus, in which nearly every chemical process is going on. It has been proved that acid and alkaline principles are constantly acting upon each other through the tissues of the animal frame ; and we have curious phenomena of endosmose and exosmose in constant action, and catalysis operating in a mysterious manner.([260])

With the refined physiological questions connected with the phenomena of sensation we cannot deal, nor will any argument be adduced for or against the hypothesis which would refer these phenomena to some extraordinary development of electric force in the brain. The entire subject appears to stand beyond the true limits of science, and every attempt to pass it is invariably found to lead to a confused mysticism, in which the real and the ideal are strangely confounded. Science stops short of the phenomena of vital action.

We cannot, however, but refer to the idea entertained by many that the brain is an electric battery, and the nerves a system of conductors. On this view Sir John Herschel remarks: "If the brain be an electric pile constantly in action, it may be conceived to discharge itself at regular intervals, when the tension of the electricity reaches a certain point, along the nerves which communicate with the heart, and thus excite the pulsation of that organ." Priestley, however, appears to have been the first to promulgate this idea.

Light is an essential element in producing the grand phenomenon of life, though its action is ill understood. Where there is light there is life, and any deprivation of this principle is rapidly followed by disease of the animal frame, and the destruction of the mental faculties. We have proof of this in the squalor of those whose necessities compel them to labour in places to which the blessings of sunshine never penetrate, as in our coal-mines, where men having everything necessary for health, except light, exhibit a singularly unhealthy appearance. The state of fatuity and wretchedness to which those individuals have been reduced, who have been subjected for years to incarceration in dark dungeons, may be referred to the same deprivation. Again, in the peculiar aspect of those people who inhabit different regions of the earth under varying influences of light, we see evidence of the powerful effects of solar action. Other forces, as yet undiscovered, may, in all probability do, exert decided influences on the animal economy; but, although we recognize many effects which we cannot refer to any known causes, we are perfectly unable to imagine the sources from which they spring.

It will be interesting now to examine the sources of animal heat, the consideration of which naturally leads us to consider the digestive system, the circulatory processes, and the effects of nervous excitation.

The theory, which attributes animal heat to the combination of the carbon of the food taken into the stomach with the oxygen of the air inspired through the lungs, has become a very favorite one. It must, however, be remembered that it is by no means a new one. The doctrines of Brown, known as the Brunonian system, and set forth in his *Elementa Medicinæ*, are founded upon similar hasty generalizations. Although, without doubt, true in a certain degree, it is not so to the extent to which its advocates would have us believe. That the carbonaceous matter received into the stomach, after having undergone the process of digestion, enters into combination with the oxygen breathed through the lungs or absorbed by the skin, and is given off from the body in the form of carbonic acid, and that, during the combination, heat is produced, by a process similar to that of ordinary combustion, is an established fact; but the idea of referring animal heat entirely to this chemical source, when there are other well-known causes producing calorific effects, is an example of the errors into which an ingenious mind may be led, when eagerly seeking to establish a favourite hypothesis.

Animal and vegetable diet, which is composed largely of carbon and hydrogen, passes into the digestive system, and becomes converted into the various matters required for the support of the animal structure. The blood is the principal fluid employed in distributing over the system the necessary elements of health and vigour, and for restoring the waste of the body. This fluid, in passing through the lungs, undergoes a very remarkable change, and not merely assumes a different colour, but really acquires new properties, from its exposure to the air with which the cells of these organs are filled. By a true chemical process, the oxygen is separated from the air, that oxygen is made to combine with the carbon and hydrogen, and carbonic acid and water are formed. These are liberated and thrown off from the body either through the lungs or by the skin. In the processes of life, as far as we are enabled to trace them, we see actions going on which are referred to certain causes which we *appear* to explain. Thus, the combination of the oxygen of the air with the carbon of the blood is truly designated a case of chemical affinity; and we find that, in endeavouring to imitate the processes of nature in the laboratory, we are, to a certain extent, successful. We can combine carbon and oxygen to produce carbonic acid; and we know that the result of that combination is the development of certain definite quantities of heat. Let us examine the conditions of this chemical phenomenon, and we shall find that in the natural and artificial processes, — for we must be allowed to make that distinction, — there are analogous circumstances. If we place a piece of pure carbon, a lump of charcoal or a diamond, in a vessel of air, or even of pure oxygen gas, no change will take place in either of these elements, and, however long they may be kept together, they will still be found as carbon or diamond, and oxygen gas. If we apply heat to the carbon until it becomes incandescent, it immediately begins to combine with the oxygen gas, — it burns; — after a little time all the carbon has disappeared, and we shall find, if the experiment has been properly made, that a gas is left behind which is distinguished by properties in every respect the reverse of those of oxygen, supporting neither life nor combustion, whereas oxygen gives increased vigour to both. We have now, indeed, carbonic acid gas formed by the union of the two principles.

A dead mass of animal matter may be placed in oxygen gas, and, unless some peculiar conditions are in some way brought about, no change will take place; but, if it were possible to apply the spark of life to it, as we light up the spark in the other case, or if, as that is beyond the power of man, we substitute a living creature, a combination between the carbon of the animal and the gas will immediately begin, and carbonic acid will be formed by the waste of animal matter, as in the other case it is by the destruction of the carbon; and, if there is not a fresh supply given, the animal must die, from the exhaustion of its fabric. Now, in both these cases, it is clear that, although this chemical union is a proximate cause of animal heat, there must be existing some power superior to it, as the ultimate cause thereof.

The slow combustion (eremacausis) of vegetable matter, decomposing under the influence of moisture and the air, does not present similar conditions to those of the human body, although it has been insisted upon to be in every respect analogous. That the results resemble each other is true, but we must carefully distinguish between effects and causes; and the results of chemical decomposition in inert matter differ from those in the living organism. The vegetable matter has lost the principle of organic life, and, that gone, the tendency of all things being to be resolved into their most simple forms, a disunion of the elements commences: oxygen and hydrogen escape from the carbon, and pass off either in the gaseous state or as water, whilst the carbon is liberated in a very finely-divided condition, and enters slowly into combination with oxygen supplied by the water or the air. Hydrogenous compounds are at the same time formed, and, under all these circumstances, as in all other chemical phenomena, an alteration of temperature results.

The animal tissue may act in the same way as platina has already been shown to act in producing combination between gases; but of this we have no proof. We know that electricity is capable of producing the required conditions, and we also learn, from the beautiful researches of Faraday, that the quantity of electricity developed during decomposition, is exactly equal to that required to effect the combination of the same elements. Thus it is quite clear that, during the combination of the carbon of the blood with the oxygen of the air, a large amount of electricity must become latent in the compound. The source of this we know not: it may be derived from some secret spring within the living structure, or it may be gathered from the matter surrounding it. There is much in nervous excitation which appears like electrical phenomena, and attempts have been frequently made to refer sensation to the agency of electricity. But these are the dreams of the ingenious, for which there is but little waking reality.

Every mechanical movement of the body occasions the development of heat; every exertion of the muscles produces sensible warmth; and, indeed, it can be shown by experiment that every expansion of muscular fibre is attended with the escape of caloric, and its contraction with the absorption of it. There are few operations of the mind which do not excite the latent caloric of the body, and frequently we find it manifested in a very remarkable manner by a suddenly-awakened feeling. The poet in the pleasure of creation glows with the ardor of his mind, and the blush of the innocent is but the exhibition of the phenomenon under some nervous excitation, produced by a spirit-disturbing thought. Thus we see that the processes of digestion and respiration are not the only sources of animal heat, but that many others exist, to which much of the natural temperature of the body must be referred.

So much that is mysterious belongs to the phenomena of life, that superstition has had a wide scope for the exercise of its influence; and through all ages a powerful party of mankind have imagined that the spirit of human curiosity must be checked before it advances to remove the veil from any physiological causes. Hence it is that even at the present day so much that stands between what, in our ignorance, we call the real and the supernatural, remains uninvestigated. Even those men whose minds are skeptical upon any development of the truths of great natural phenomena, — who, at all events, will have proof before they admit the evidence, are ready to give credit to the grossest absurdities which may be palmed upon them by ingenious charletans, where the subject is man and his relations to the spiritual world.

Man, and the races of animals by which he is surrounded, present a very striking group, consider them in whatever light we please. The gradual improvement of organic form, and the consequent increase of sensibility, and eventually the development of reason, are the grandest features of animated creation. The conditions as to number even of the various classes are not the least remarkable phenomena of life. In the lowest orders of animals, creatures of imperfect organization, — consequently those to whom the conditions of pain must be nearly unknown, — increase by countless myriads. Of the infusoria and other beings, entire mountains have been formed, although microscopes of the highest powers are required to detect an individual. Higher in the scale, even among insects, the same remarkable conditions of increase are observed. Some silkworms lay from 1,000 to 2,000 eggs; the wasp deposits 3,000; the ant from 4,000 to 5,000. The queen bee lays between 5,000 and 6,000 eggs, according to Burmeister; but Kirby and Spence state that in one season the number may amount to 40,000 or 50,000. But, above all, the white ant *(Termes fatalis)* produces 86,400 eggs each day, which, continuing for a lunar month, gives the astonishing number of 2,419,200, a number far exceeding that produced by any known animal.

These may appear like the statements in which a fictionist might indulge, but they are the sober truths discovered by the most pains-taking and cautious observers. And it is necessary that such conditions should prevail. These insects, and all the lower tribes of the animal kingdom, furnish food for the more elevated races. Thousands are born in an hour, and millions upon millions perish in a day. For the support of organic life, like matter is required; and we find that the creatures who are destined to become the prey of others, are so constituted that they pass from life with a perfect unconsciousness of suffering. As the animal creation advances in size and strength, their increase becomes limited; and thus they are prevented from maintaining by numbers that dominion over the world which they would be enabled from their powers to do, were their bands more numerous than we now find them.

The comparative strength, too, of the insect tribes has ever been a subject of wonder and of admiration to the naturalist. The strength of these minute creatures is enormous; their muscular power, in relation to their size, far exceeds that of any other animal. The grasshopper, will spring two hundred times the length of its own body. The dragon-fly, by its strength of wing, will sustain itself in the air for a long summer day with unabated speed. The house-fly makes six hundred strokes with its wings, which will carry it five feet, every second.

Such are the wonders of the natural world; from the zoophyte, growing like a flowering plant([261]) upon an axis filled with living pith — a small remove from the conditions of vegetable life, upwards through the myriads of breathing things — to man, we see the dependence of all upon these physical powers which we have been considering.

To trace the effects of these great causes through all their mysterious phases is the work of inductive science; and the truths discovered tend to fit us for the enjoyment of the eternal state of high intelligence to which every human soul aspires.

That which the ignorant man calls the supernatural, the philosopher classes amongst natural phenomena. The ideal of the credulous man becomes the real to one who will bend his mind to the task of inquiry. Therefore, to attempt to advance our knowledge of the unknown, to add to the stores of truth, is an employment worthy the high destiny of the human race. Remembering that the revelations of natural science cannot in any way injure the revelation of eternal truth, but, on the contrary, aid to establish in the minds of the doubting a firm conviction of its Divine origin and of man's high position, we need never fear that we are proceeding too far with any inquiry so long as we are cautious to examine the conditions of our own minds, that they may not be made the dupe of the senses.

In the fairies of the hills and valleys, in the gnomes of the caverns, in the spirits of the elements, we have the attempts of the mind, when the world was young, to give form to the dim outshadowings of something which was then felt to be hidden behind external nature.

In the Oread, the Dryad, and the Nereid, we have, in like manner, an embodiment of powers which the poet-philosopher saw in his visions presiding over the mountain, the forest, and the ocean. Content with these, invested as they were with poetic beauty, man for ages held them most religiously sacred ; but the progress of natural science has destroyed this class of creations. "Great Pan is dead," but the mountains are not voiceless ; they speak in a more convincing tone ; and, instead of the ear catching the dying echo of an obscure truth, it is gladdened with the full, clear note of Nature, in the sweetest voice, proclaiming secrets which were unknown to the dreams of superstition.

CHAPTER XVI.

GENERAL CONCLUSIONS.

The Changes produced on Physical Phenomena by the Movement of the Solar System considered — Exertion of the Physical Forces through the Celestial Spaces — The Balance of Powers — Varieties of Matter — Extension of Matter — Theory of Nonentity — A Material Creation, an indisputable Fact — Advantages of the Study of Science — Conclusion.

We have examined terrestrial phenomena under many of the harmonious conditions which, with our limited intelligence, we can reach by the aid of science. From the first exhibition of force, in the cohesion of two atoms, onward to the full development of organic form in the highest order of animals, we have observed strange influences. We have seen the solitary molecule invested with peculiar properties, and regulated by mighty forces; we have learned that the modes of motion given to this beautiful sphere produce curious changes in the operation of these powers; and we may with safety infer that every atom constituting this globe is held in wonderful suspension against every atom of every star, in the celestial spaces, even to that bright orb in the centre of the Pleiades, around which the entire system of created worlds is supposed to roll.

As we move around our own sun — in the limited period of 365 days — we experience transitions from heat to cold, dependent upon our position in regard to that luminary. May we not therefore conclude, without being charged with making any violent deduction, that, in the great revolution of our system around the centre of space, we are undergoing gradual changes which are essential to the great scheme of creation, though at present incomprehensible to us?

In our consideration of the influence of time on the structure of the earth as we find it, we discover that, in ages long past, the vegetation of the tropics existed upon these northern parts of the globe; and geological research has also proved that over the same lands the cold of an arctic winter must have long prevailed — the immense glaciers of that period having left the marks of their movements upon the face of the existing rocks. ([262]) We know that during 3,000 years no change of temperature has taken place in the European climate. The children of Israel found the date and the vine flourishing in Canaan; and they exist there still. Arago has shown that a trifling alteration of temperature would have destroyed one or the other of these fruit-bearing trees, since the vine will not ripen where the mean temperature of the year is higher than 84°, or the date flourish where it sinks below that degree.

How immense, then, the duration of time since these changes must have taken place! The 432,000 years of Oriental mythology is a period scarcely commensurable with these effects; yet, to the creature of threescore years, that period appears an eternity. The thirty-three millions of geographical miles which our solar system traverses annually, if multiplied by three thousand years, during which we know no change has taken place, give us 99,000,000,000 as the distance passed over in that period. How wide, then, must have been the journey of the system in space to produce the alteration in the physical powers, by which these changes have been effected!

We have an example, and a striking one, of the variations which may be produced in all the physical conditions of a world, in those disturbances of Uranus which led to the discovery of Neptune. For thirty years or more certain perturbations were observed in this distant planet, the discovery of Sir William Herschel, and calculation pointed to some still more remote mass of matter as the cause, which has been verified by its actual discovery. But now Uranus is at rest; — quietly that star progresses in its appointed orbit, — Neptune can no longer cause it to rock upon its centre, — they are too remote to produce any sensible influence upon each other. Consequently, for thir y years, it is evident, phenomena must have occurred on the surface of Uranus, which can be no longer repeated until these two planets again arrive at the same positions in their respective paths which they have occupied since 1812. These considerations assist us in our attempts to comprehend infinite time and space; but the human mind fails to advance far in the great sublimity.

Through every inch of space we have evidence of the exercise of such forces as we have been considering. Gravitation chains world to world, and holds them all suspended from the mystic centre. Cohesion binds every mass of matter into a sphere. Heat, radiating from one planet to another, does its work in all, giving variety to matter. Light seeks out every world — each trembling star tells of the mystery of its presence. Where light and heat are, chemical action, as an associated power, must be present; and electricity must do its wondrous duties amongst them all. Modified by peculiar properties of matter, they may not manifest themselves in phenomena like those of our terrestrial nature; but the evidence of light is a sufficient proof of the presence of its kindred elements; and it is difficult to imagine all these powers in action without producing some form of organization. In the rounded pebble which we gather from the seashore, — in the medusa floating bright with all the beauty of prismatic colour in the sunlit sea, — in the animal, mighty in his strength, roaming the labyrinthine forests, or, great in intelligence, looking from this to the mysteries of other worlds,

— in all created things around us, we see direct evidence of a beautiful adjustment of the balance of forces, and the harmonious arrangement of properties.

One atom is removed from a mass and its character is changed; one force being rendered more active than another, and the body, under its influence, ceases to be the same in condition. The regulation which disposes the arrangements of matter on this earth, must exist through the celestial spaces, and every planet bears the same relation to every other glittering mass in heaven's o'erarching canopy, as one atom bears to another in the pebble, the medusa, the lion, or the man. An indissoluble bond unites them all, and the grain of sand which lies buried in the depth of one of our primary formations holds, chained to it by these all-pervading forces, the uncounted worlds which, like luminous sand, are sprinkled by the hand of the Creator through the universe. Thus we advance to a conception of the oneness of creation.

The vigorous mind of that immortal bard who sang "of man's first disobedience," never, in the highest rapture, the holiest trance of poetic conception, dreamed of any natural truths so sublime as those which science has revealed to us.

The dependence of all the systems of worlds upon each other, every dust composing each individual globe being "weighed in a balance," the adjustment of the powers by which every physical condition is ordered, the disposition of matter in the mass of the earth, and the close relation of the kingdoms of nature, — are all revelations of natural truths, exalting the mind to the divine conception of the universe.

There is a remarkable antagonism displayed in the operation of many of these forces. Gravitation and cohesion act in opposition to the repellent influences of caloric. Light and heat are often associated in a very remarkable manner; but they are certainly in their radiant states in antagonism to chemical action, whether produced by the direct agency of actinic force, or through the intermediate excitement of the electrical current. ([263]) And in relation to chemical force, as manifested in organic combinations, we have the all-powerful operation of LIFE preventing any exercise of its decomposing power.([264]) As world is balanced against world in the universe, so in the human fabric, in the vegetable structure, in the crystallized gem, or in the rude rock, force is weighed against force, and the balance hangs in tranquillity. Let but a slight disturbance occasion a vibration of the beam, and electricity shakes the stoutest heart with terror, at the might of its devastating power.([265]) Heat melts the hardest rocks, and the earth trembles with volcanic strugglings; and actinic agency, being freed from its chains, speedily spreads decay over the beautiful, and renders the lovely repulsive.

We know matter in an infinite variety of forms, from the most ponderous metal to the lightest gas; and we have it within our power to render the most solid bodies invisible in the condition of vapour. Is it not easy, then, to understand that matter may exist equally attenuated in relation to hydrogen, as that gas itself is, when compared with the metal platinum? A doubt has been raised against this view, from the difficulty of accounting for the passage of the physical elements through solid masses of matter. If we, however, remember that the known gases have the power of transpiration through matter in a remarkable degree,([266]) and that the passage of water through a sieve may be prevented by heat, it will be at once apparent that the permeation of any radiant body through fixed solid matter, is entirely a question of conditions.

We can form no idea of the size of the ultimate atom; we cannot comprehend the degree of etherealization to which matter may be extended. Our atmosphere, we have seen, is only another condition of the same elements which compose all the organized forms of matter upon the earth, and, at the height reached by man, it is in a state of extreme attenuation. What must be its condition at the distance of forty miles from the earth? According to known laws, certain phenomena of refraction have led us to set these bounds to the matter constituting our globe: but it may exist in such a state of tenuity, that no philosophical instrument constructed by human hands could measure its refracting power; and who shall declare with certainty that matter itself may not be as far extended as we suppose its influences to be?

"Hast thou perceived the breadth of the earth? declare, if thou knowest it all.

"Knowest thou the ordinances of heaven? Canst thou set the dominion thereof in the earth?"

A cheerless philosophy would teach us to regard all things as the mere exhibition of properties, a manifestation of powers; it believes not in a material creation. The grandeur of the earth, and the beautiful forms adorning it, are not entities. Yonder exquisite specimen of the skill of man, in which mind appears to shine through the marble, — that distant mountain which divides the clouds as they are driven by the winds across it, — those trees, amid whose branches the birds make most melodious music, — this flower, so redolent of perfume, so bright in colour, and so symmetric in form, — and that lovely being who, a model of beauty and grace, walks the earth an impersonation of love and charity blended, making, indeed, "a sunshine in a shady place," are not realities. Certain forces combine to produce effects, all of which unite to deceive poor man into the belief that he is a material being, and the inhabitant of a material world. There may be ingenuity in the philosophy of this school; its metaphysics may be of a high order; but it evidently advances from the real to the ideal with such rapidity, that every argument is based on an assumption without a proof; every assumption being merely a type of the philosophy itself, — a baseless fabric, a transcendental vision.

A material creation surrounds us. This earth, and all that it contains, and the immense hosts of stellar worlds, are absolute entities, surrounded with, and interpenetrated by, certain exhibitions of creative intelligence, which perform, according to fixed laws, the mighty labours upon which depend the infinite and eternal mutations of matter. The origin of a grain of dust is hidden from our finite comprehensions; but its existence should be a source of hope, that those minds which are allowed the privilege of tracing out its marvellous properties, — of examining the empyreal principles upon which its condition, as a grain of dust, depends, — and even of reducing these giant elements to do our human bidding, — may, after a period of probation, be admitted to the enjoyment of that infinite power to which the great secrets of creation will be unveiled.

Every motion which the accurate search of the experimentalist has traced, every principle of power which the physicist has discovered, every combination which the chemist has detected, every form which the naturalist has recorded, involves reflections of an exalting character, which constitute the elements of the highest poetry. The philosophy of physical science is a grand epic, the record of natural science a great didactic poem.

To study science for its useful applications merely, is to limit its advantages to purely sensual ends. To pursue science for the sake of the truths it may reveal, is an endeavour to advance the elements of human happiness through the intelligence of the race. To avail ourselves of facts for the improvement of art and manufactures, is the duty of every nation moving in the advance of civilization. But to draw from the great truths of science intelligible inferences and masterly deductions, and from these to advance to new and beautiful abstractions, is a mental exercise which tends to the refinement and elevation of every human feeling.

The mind thus exercised during the mid-day of life, will find in the twilight of age a divine serenity; and, charmed by the music of nature, which like a vesper hymn poured forth from pious souls, proclaims in devotion's purest strain the departure of day, he will sink into the repose of that mysterious night which awaits us all, tranquil in the happy consciousness that the sun of truth will rise in unclouded brilliancy, and place him in the enjoyment of that intellectual light, which has ever been among the holiest aspirations of the human race.

The task of wielding the wand of science, — of standing a scientific evocator within the charmed circle of its powers, is one which leads the mind through nature up to nature's God.

Experiment and observation instruct us in the discovery of a fact; — that fact connects itself with natural phenomena, — the ultimate cause of which we learn from Divine revelation, and receive in full belief, — but the proximate causes are reserved as trials of man's intelligence; and every natural truth, discovered by induction, enables the contemplative mind to deduce those perfect laws which are exemplifications of the fresh-springing and all-enduring Poetry of Science.

NOTES.

(1) p. 28. — The view entertained by Dr. Faraday, which will be comprehended from one or two short extracts from his valuable and suggestive paper, claims attention: —

"If the view of the constitution of matter already referred to be assumed to be correct — and I may be allowed to speak of the particles of matter, and the space between them (in water, or in the vapour of water, for instance,) as two different things — the space must be taken as the only continuous part, for the particles are considered as separated by space from each other. Space will permeate all masses of matter in every direction like a net, except that, in the place of meshes, it will form cells, isolating each atom from its neighbours, and itself only being continuous."

Examining the question of the conducting power of different bodies, and observing that as space is the only continuous part, so space, according to the received view of matter, must be at one time a conductor, at others a non-conductor, it is remarked: —

"It would seem, therefore, that, in accepting the ordinary atomic theory, space may be proved to be a non-conductor in non-conducting bodies, and a conductor in conducting bodies; but the reasoning ends in this — a subversion of that theory altogether; for if space be an insulator, it cannot exist in conducting bodies; and if it be a conductor, it cannot exist in insulating bodies." — *A Speculation touching Electric Conduction, and the Nature of Matter*: by Michael Faraday, D. C. L., F. R. S., &c.: Philosophical Magazine, vol. xxiv. Third Series.

See also Wollaston, *On the Finite Extent of the Atmosphere.* — Phil. Trans. 1822.

——— Young, *On the Essential Properties of Matter.*—Lectures on Natural Philosophy.

——— Mossotti, *On Molecular Action.* — Scientific Memoirs, vol. i. p. 448.

(2) p. 30. — "Motion, therefore, is a change of rectilinear distance between two points. Allowing the accuracy of this definition, it appears that two points are necessary to constitute motion; that in all cases, when we are inquiring whether or not any body or point is in motion, we must recur to some other point which we can compare with it; and that, if a single atom existed alone in the universe, it could neither be said to be in motion nor at rest.

"The space which we call quiescent is in general the earth's surface; yet we well know, from astronomical considerations, that every point of the earth's surface is perpetually in motion, and that in very various directions: nor are any material objects accessible to our senses which we can consider as absolutely motionless, or even as motionless with regard to each other; since the continual variation of temperature to which all bodies are liable, and the minute agitations arising from the motion of other bodies with which they are connected, will always tend to produce some imperceptible changes in their distances." — *Lectures on Natural Philosophy, &c.*, by Thomas Young, M. D. Edited by Rev. P. Kelland. 1845.

(3) p. 30. — "The position which I seek to establish in this essay is, that the various imponderable agencies, or the affections of matter which constitute the main objects of experimental physics, viz. heat, light, electricity, magnetism, chemical affinity, and motion, are all correlative, or have a reciprocal dependence; — that neither, taken abstractedly, can be said to be the essential or proximate cause of the others; but that either may, as a force, produce, or be convertible into, the other: — thus heat may mediately or immediately produce electricity, electricity may produce heat, and so of the rest. . . . Although strongly inclined to believe that the five other affections of matter, which I have above named, are, and will ultimately be, resolved into modes of motion, it would be going too far at present to assume their identity with it: I, therefore, use the term force, in reference to them, as meaning that active force inseparable from matter, which induces its various changes." — *On the Correlation of Physical Forces:* by W. R. Grove, Esq., M. A., F. R. S.

(4) p. 30. — When discussing the hypothesis of Hobbes — *that no body can possibly be moved but by a body contiguous and moved* — Boyle asks: —

"I demand how there comes to be local motion in the world? For either all the portions of matter that compose the universe have motion belonging to their natures, which the Epicureans affirm for their atoms, or some parts of matter have this motive power, and some have not, or else none of them have it; but all of them are naturally devoid of motion. If it be granted that motion does naturally belong to all parts of matter, the dispute is at an end, the concession quite overthrowing the hypothesis.

"If Mr. Hobbes should reply that the motion is impressed upon any of the parts of matter by God, he will say that which I most readily grant to be true, but will not serve his turn, if he would speak congruously with his own hypothesis. For I demand whether this Supreme Being that the assertion has recourse to, be a corporeal or an incorporeal substance? If it be the latter, and yet the efficient cause of motion in bodies, then it will not be universally true that whatever body is moved is so by a body contiguous and moved. For, in our supposition, the bodies that God moves, either immediately or by the intervention of any other immaterial being, are not moved by a body contiguous, but by an incorporeal spirit." — *Some Considerations about the Reconcileableness of Reason and Religion:* Boyle, vol. iii. p. 520.

([5]) p. 31. — Boyle has some ingenious speculations on this point: —

"That there is local motion in many parts of matter is manifest to sense, but how matter came by this motion was of old, and is still, hotly disputed of: for the ancient Corpuscularian philosophers (whose doctrine in most other points, though not in all, we are the most inclineable to), not acknowledging an author of the universe, were thereby reduced to make motion congenite to matter, and consequently coeval with it. But since local motion, or an endeavour at it, is not included in the nature of matter, which is as much matter when it rests as when it moves; and since we see that the same portion of matter may from motion be reduced to rest, and after it hath continued at rest, so long as other bodies do not put it out of that state, may by external agents be set a moving again; I, who am not wont to think a man the worse naturalist for not being an atheist, shall not scruple to say with an eminent philosopher of old, whom I find to have proposed among the Greeks that opinion (for the main) that the excellent Des Cartes has revived amongst us, that the origin of motion in matter is from God; and not only so, but that thinking it very unfit to be believed, that matter barely put into motion, and then left to itself, should casually constitute this beautiful and orderly world; I think also further, that the wise Author of things did, by establishing the laws of motion among bodies, and by guiding the first motions of the small parts of matter, bring them to convene after the manner requisite to compose the world; and especially did contrive those curious and elaborate engines, the bodies of living creatures, endowing most of them with the power of propagating their species." — *Considerations and Experiments touching the Origin of Forms and Qualities:* Boyle's Works, vol. ii. p. 460. Edinburgh. 1744.

(6) p. 31. — Cudworth's *Intellectual System.*

(7) p. 9. — "According to the Pythagoreans and Platonists, there is a life infused throughout all things . . . . an intellectual and artificial fire — an inward principle, animal spirit, or natural life, producing or forming within, as art doth without — regulating, moderating, and reconciling the various motions, qualities, and parts of the mundane system. By virtue of this life, the great masses are held together in their ordinary courses, as well as the minutest particles governed in their natural motions, according to the several laws of attraction, gravity, electricity, magnetism, and the rest. It is this gives instincts, teaches the spider her web, and the bee her honey; — this it is that directs the roots of plants to draw forth juice from the earth, and the leaves and the cortical vessels to separate and attract such particles of air and elementary fire as suit their respective natures." — Bishop Berkeley. *Siris*, No. 27 7.

(8) p. 31. — "The revolution of the earth is performed in a natural day, or, more strictly speaking, once in 23h. 56′ 4″, and as its mean circumference is 24,871 miles, it follows that any point in its equatorial surface has a rotary motion of more than 1,000 miles per hour. This velocity must gradually diminish to nothing at either pole. Whilst the earth is thus revolving on its axis, it has a progressive motion in its orbit. If we take the length of the earth's orbit at 630,000,000, its motion through space must exceed 68,490 miles in the hour." — Enc. Brit. *Art. Physical Geography.*

(9) p. 32. — "Here then we have the splendid result of the united studies of MM. Argelander, O. Struve, and Peters, grounded on observations made at the three observatories of Dorpat, Abo, Pulkova, and which is expressed in the following thesis: The motion of the solar system in space is directed to a point of the celestial vault situated on the right line which joins the two stars π and μ *Herculis*, at a quarter of the apparent distance of these stars, reckoning from π *Herculis*. The velocity of this motion is such that the sun, with all the bodies which depend upon it, advances annually in the above direction 1.623 times the radius of the earth's orbit, or 33,550,000 geographical miles. The possible error of this last number amounts to 1,733,000 geographical

miles, or to a *seventh* of the whole value. We may then wager 400,000 to 1 that the sun has a proper progressive motion, and 1 to 1 that it is comprised between the limits of thirty-eight and twenty-nine millions of geographical miles." — *Etudes d'Astronomie Stellaire; Sur la Voie Lactée et sur les Distances des Etoiles Fixes:* M. F. W. G. Struve. [A report addressed to his Excellency M. Le Comte Ouvaroff, Minister of Public Instruction and President of the Imperial Academy of Sciences at St. Petersburg.]

([10]) p. 33. — "The first great agent which the analysis of natural phenomena offers to our consideration, more frequently and prominently than any other, is force. Its effects are either, 1st, to counteract the exertion of opposing force, and thereby to maintain *equilibrium;* or, 2ndly, to produce *motion* in matter.

"Matter, or that whatever it be of which all the objects in nature which manifest themselves directly to our senses consist, presents us with two general qualities — which at first sight appear to stand in contradiction to each other — activity and inertness. Its activity is proved by its power of spontaneously setting other matter in motion, and of itself obeying their mutual impulse, and moving under the influence of its own and other force; inertness, in refusing to move unless obliged to do so by a force impressed externally, or mutually exerted between itself and other matter, and by persisting in its state of motion or rest unless disturbed by some external cause. Yet, in reality, this contradiction is only apparent. Force being the cause, and motion the effect produced by it on matter, to say that matter is inert, or has *inertia,* as it is termed, is only to say that the cause is expended in producing its effect, and that the same cause cannot (without renewal) produce double or triple its own proper effect. In this point of view, equilibrium may be conceived as a continual production of two opposite effects, each undoing at every instant what the other has done." — See continuation of the argument, in *Herschel's Discourse on the study of Natural Philosophy,* page 223.

In the Edinburgh New Philosophical Journal, vol. xlv., will be found a paper by Dr. Robert Brown — "*Of the sources of motions upon the Earth, and of the means by which they are sustained,*" which will well repay an attentive perusal, as pointing to a class of investigation of the highest order, and containing deductions of the most philosophic description.

([11]) p. 34. — Friction, it is well known, generates heat: by rapidly rubbing two sticks together, the Indian produces their ignition; heat and light being both manifested. Under every mechanical disturbance electrical changes can be detected, and the action of heat in the combustion of the wood is a chemical phenomenon.

([12]) p. 35. — Count Rumford's experiment consisted in placing a mass of metal in a box of water at a known temperature, and, by employing a boring apparatus, ascertaining carefully the increase of heat after a given number of revolutions. He thus describes his most satisfactory experiment: —

"Everything being ready, I proceeded to make the experiment I had projected, in the following manner: The hollow cylinder having been previously cleaned out, and the inside of its bore wiped with a clean towel till it was quite dry, the square iron bar, with the blunt steel borer fixed to the end of it, was put into its place; the mouth of the bore of the cylinder being closed at the same time by means of the circular piston through the centre of which the iron bar passed.

"This being done, the box was put in its place; and the joinings of the iron rod, and of the neck of the cylinder with the two ends of the box, having been made water-tight, by means of collars of oiled leather, the box was filled with cold water (viz., at the temperature of 60) and the machine was put in motion. The result of this beautiful experiment was very striking, and the pleasure it afforded me amply repaid me for all the trouble I had had, in contriving and arranging the complicated machinery used in making it. The cylinder, revolving at the rate of about thirty-two times in a minute, had been in motion but a short time, when I perceived, by putting my hand into the water and touching the outside of the cylinder, that heat was generated, and it was not long before the water which surrounded the cylinder began to be sensibly warm. At the end of one hour, I found, by plunging a thermometer into the water in the box (the quantity of which fluid amounted to 18.77 lbs. avoirdupois, or 2¼ wine gallons), that its temperature had been raised no less than 47°; being now 107° of Fahrenheit's scale. When thirty minutes more had elapsed, or one hour and thirty minutes after the machinery had been put in motion, the heat of the water in the box was 142°. At the end of two hours, reckoning from the beginning of the experiment, the temperature of the water was found to be raised to 178°. At two hours twenty minutes, it was 200°; and at two hours thirty minutes it *actually boiled*." — *Inquiry concerning the Source of the Heat excited by Friction*: Philosophical Transactions, vol. lxxxviii. A.D. 1798.

Mr. Joule brought a communication on the same subject before the British Association at Cambridge, which was afterwards published in the Philosophical Magazine, and from that journal the following notices are extracted: —

"The apparatus exhibited before the Association consisted of a brass paddle-wheel, working horizontally in a can of water. Motion could be communicated to this paddle by means of weights, pulleys, &c. The paddle moved with great resistance in the can of water, so that the weights (each of four pounds) descended at the slow rate of about one foot per second. The height of the pulleys from the ground was twelve yards, and consequently when the weights had descended through that distance they had to be wound up again in order to renew the motion of the paddle. After this operation had been repeated sixteen times, the increase of the temperature of the water was ascertained by means of a very sensible and accurate thermometer.

"A series of nine experiments was performed in the above manner, and nine experiments were made in order to eliminate the cooling or heating effects of the atmosphere. After reducing the result to the capacity for heat of a pound of water, it appeared that for each degree of heat evolved by the friction of water, a mechanical power equal to that which can raise a weight of 890 lbs. to the height of one foot, had been expended.

"Any of your readers who are so fortunate as to reside amid the romantic scenery of Wales or Scotland could, I doubt not, confirm my experiments by trying the temperature of the water at the top and at the bottom of a cascade. If my views be correct, a fall of 817 feet will of course generate one degree of heat, and the temperature of the river Niagara will be raised about one-fifth of a degree by its fall of 160 feet." — *Relation between Heat and Mechanical Power;* Philosoph. Mag. vol. xxvii. 1845.

(13) p. 36. — Three hypotheses may be used to account for this most curious phenomenon.

1st. The body shines by its own light and then explodes like a sky-rocket, breaking into minute fragments too small to be any longer visible to the naked eye.

2nd. Such a body, having shone by its own light, suddenly ceases to be luminous. "The falling stars and other fiery meteors which are frequently seen at a considerable height in the atmosphere, and which have received different names according to the variety of their figure and size, arise from the fermentation of the effluvia of acid and alkaline bodies which float in the atmosphere. When the more subtile parts of the effluvia are burned away, the viscous and earthy parts become too heavy for the air to support, and by their gravity fall to the earth." —Keith's *Use of the Globes*. According to Sir Humphrey Davy, in the Philosophical Transactions for 1817, "the luminous appearances of shooting stars and meteors cannot be owing to any inflammation of elastic fluids, but must depend upon the ignition of solid bodies."

3. The body shines by the reflected light of the sun, and ceases to be visible by its passing into the earth's shadow, or, in other words, is eclipsed. Upon the two former suppositions, the fact of the star's disappearance conveys to us no knowledge of its position, or of its distance from the earth; and all that can be said, is, that if it be a satellite of the earth, the great rapidity of its motion involves the necessity of its being at no great distance from the earth's surface—much nearer than the moon; while the resistance it would encounter in traversing the air would be so great that it is probably without the limits of our atmosphere. *Sir J. W. Lubbock leans to the third hypothesis.*—Sir J. W. Lubbock, *On Shooting Stars*: Phil. Mag. No. 213, p. 81.

Sir J. Lubbock also published a supplementary paper on the same subject, in No. 214, p. 170.

Mr. J. P. Joule entertains an hypothesis with respect to Shooting Stars similar to that advocated by Chladni to account for meteoric stones, and he reckons the ignition of these miniature planetary bodies by their violent collision with our atmosphere, to be a remarkable illustration of the doctrine of the equivalency of heat to mechanical power, or *vis viva*.

If we suppose a meteoric stone of the size of a six-inch cube to enter our atmosphere at the rate of eighteen miles per second of time, the atmosphere being $\frac{1}{100}$ of its density at the earth's surface, the resistance offered to the motion of the stone will in this case be at least 51,600 lbs.; and if the stone traverse twenty miles with this amount of resistance, sufficient heat will thereby be developed to give 1° Fahrenheit to 6,967,980 lbs. of water. Of course by far the largest portion of this heat will be given to the displaced air, every particle of which will sustain the shock, whilst only the surface of the stone will be in violent collision with the atmosphere. Hence the stone may be considered as placed in a blast of intensely heated air, the heat being communicated from the surface to the centre by conduction. Only a small portion of the heat evolved will therefore be received by the stone; but if we estimate it at only $\frac{1}{100}$, it will still be equal to 1° Fahrenheit per 69,679 lbs. of water, a quantity quite equal to the melting and dissipation of any materials of which it may be composed. — Mr. J. P. Joule, *On Shooting Stars;* Phil. Mag. No. 216, p. 348.

(14) p. 37. — "Laplace conjectures that in the original condition of the solar system, the sun revolved upon his axis, surrounded by an atmosphere which, in virtue of an excessive heat, extended far beyond the orbits of all the planets, the planets as yet having no existence. The heat gradually diminished, and as the solar atmosphere contracted by cooling, the rapidity of its rotation increased by the laws of rotatory motion; and an exterior zone of vapour was detached from the rest, the central attraction being no longer able to overcome the increased centrifugal force. This zone of vapour might in some cases retain its form, as we see it in Saturn's ring; but more usually the ring of vapour would break into several masses, and these would generally coalesce into one mass, which would revolve about the sun." — Whewell's *Bridgewater Treatise.*

The following passage is translated by the same author from Laplace: —

"The anterior state (a state of cloudy brightness) was itself preceded by other states, in which the nebulous matter was more and more diffuse, the nucleus being less and less luminous. We arrive in this manner at a nebulosity so diffuse, that its existence could scarce be suspected. Such is in fact the first state of the nebula which Herschel carefully observed by means of his telescope."

Sir William Herschel has the following observations on these remarkable masses: —

"The nature of planetary nebulæ, which has hitherto been involved in much darkness, may now be explained with some degree of satisfaction, since the uniform and very considerable brightness of their apparent disc accords remarkably well with a much condensed, luminous fluid; whereas, to suppose them to consist of clustering stars will not so completely account for the milkiness or soft tint of their light, to produce which it would be required that the condensation of the stars should be carried to an almost inconceivable degree of accumulation.

"How far the light that is perpetually emitted from millions of suns may be concerned in this shining fluid, it might be presumptuous to attempt to determine; but notwithstanding the inconceivable subtility of the particles of light, when the number of the emitting bodies is almost infinitely great, and the time of the continual emission indefinitely long, the quantity of emitted particles may well become adequate to the constitution of a shining fluid or luminous matter, provided a cause can be found that may retain them from *flying off, or reunite them."* — *Observations on Nebulous Stars:* Philosophical Transactions, vol. lxxxi. A.D. 1791.

In addition, the following Memoirs on the same subject, by Sir William Herschel, have been published in the Philosophical Transactions: *Catalogue of* 1,000 *Nebulæ and clusters of Stars*, vol. lxxvi.; *Catalogue of another* 1,000, *with remarks on the heavens*, vol. lxxix.; *Catalogue of* 500 *more, with remarks as above*, vol. xcii.; *Of such as have a cometary appearance*, vol. ci.; *Of planetary nebulæ*, ibid.; *Of stellar nebulæ*, ibid.; *On the sidereal part of the heavens, and its connection with the nebulous*, vol. civ.; *On the relative distances of clusters of nebulous stars*, vol. cviii.

([15]) p. 37. — "Lord Rosse's beautiful telescopes have been formed upon principles which appear to embrace the best possible conditions for obtaining a reflecting surface which should reflect the greatest quantity of light, and retain that property little diminished for a length of time. The alloy used for this purpose consists of tin and copper in atomic proportions, namely, one atom of tin to four atoms of copper, or by weight 58.9 to 126.4." — *On the Construction of large Reflecting Telescopes:* by Lord Rosse. Report of the Fourteenth Meeting of the British Association, 1854; p. 79.

([16]) p. 38. — The best description of the Zodiacal Light occurs in a letter furnished by Sir John Herschel to the *Times* newspaper in March, 1843: "The zodiacal light, as its name imports, invariably appears in the zodiac, or, to speak more precisely, in the plane of the sun's equator, which is 7° inclined to the zodiac, and which plane, seen from the sun, intersects the ecliptic in longitude 78° and 258°, or so much in advance of the equinoctial points: in consequence it is seen to the best advantage at, or a little after, the equinoxes; after sunset, at the spring, and before sunrise, at the autumnal equinox; not only because the direction of its apparent axis lies at those times more nearly perpendicular to the horizon, but also because at those epochs we are approaching the situation when it is seen most completely in section.

"At the vernal equinox, the appearance of the zodiacal light is that of a pretty broad pyramidal, or rather lenticular, body of light, which begins to be visible as soon as the twilight decays. It is very bright at its lower or broader part near the horizon, and, if there be broken clouds about, often appears like the glow of a distant conflagration, or of the rising moon, only less red; giving rise, in short, to amorphous masses of light, such as have been noticed by one of your correspondents as possibly appertaining to the comet. At higher altitudes, its light fades gradually, and is seldom traceable much beyond the Pleiades, which it usually, however, attains and involves, and (what is most to my present purpose) its axis at the vernal equinox is always inclined (to the northward of the equator) at an angle of between 60° and 70° to the horizon, and it is most luminous at its base, resting on the horizon, where also it is broadest, occupying, in fact, an angular breadth of somewhere about 10° or 12° in ordinary clear weather."

([17]) p. 38. — "The assumption that the extent of the starry firmament is literally infinite has been made by one of the greatest of astronomers, the late Dr. Olbers, the basis of a conclusion that the celestial spaces are, in some slight degree, deficient in *transparency;* so that all beyond a certain distance is, and must remain for ever, unseen; the geometrical progression of the extinction of light far outrunning the effect of any conceivable increase in the power of our telescopes. Were it not so, it is argued, every part of the celestial concave ought to shine with the brightness of the solar disc, since no visual ray could be so directed as not, in some point or other of its infinite length, to encounter such a disc." — *Edinburgh Review*, p. 185, for January, 1848; *Etudes d'Astronomie Stellaire.*

([18]) p. 39. — In the *Astronomische Nachrichten* of July, 1846, appeared a Memoir by M. Mädler, *Die Centralsonne.* The conclusions arrived at by Mädler, may be understood from the following quotation from a French translation, made by M. A. Gautier, in the *Archives des Sciences Physiques et Naturelles*, for October, 1846. — "Quoiqu'il résulte de ce qui précède que la région du ciel que j'ai adoptée satisfait à toutes les conditions posées plus haut, il n'en est pas moins convenable de la soumettre à toutes les épreuves possibles. Plusieurs essais de combinaisons différentes m'ont convaincu qu'on ne pourrait trouver aucun autre point dans le ciel qui pût tenir lieu, même d'une manière approchée, que celui que j'ai adopté. On pourrait maintenant m'addresser l'objection que, si la région du ciel où se trouve le centre de gravité de notre système d'étoiles fixes, est determinée par ce qui précède entre certaines limites, il n'en résulte pas la nécessité de choisir Alcyone pour ce centre, attendu qu'il pourrait bien tomber sur quelqu'autre étoile située dans le groupe ou dans son voisinage. Mais outre que c'est tout près de là que se trouve le groupe le plus brillant et le plus riche en étoiles de tout le ciel, et qu'il ne s'agit point ici d'un point arbitraire situé dans le voisinage peu apparent et qui n'ait rien qui le distingue, il ne se trouve nul part, même dans la région voisine, une aussi exacte concordance des mouvements propres qu'ici, et ces mouvements correspondent mieux que tous les autres aux conditions établies plus haut. Or si l'on doit considérer ce groupe central, entre les étoiles également éloignées, on peut présumer que la plus brillante de beaucoup présente la plus grande masse. Outre cela Alcyone, considérée optiquement, est au milieu du groupe des Pleïades ; et son mouvement propre, déterminé par Bessel, est plus exactement en accord avec la moyenne de ceux des autres Pleïades ; ainsi que des étoiles de cette région jusqu'à 10° de distance. *Je puis donc établir comme conséquence de tout ce qui précède que le groupe des Pléïades est le groupe central de l'ensemble du système des étoiles fixes, jusqu'aux limites extérieures déterminées par la Voie Lactée ; et que Alcyone est l'étoile de ce groupe qui paraît être, le plus vraisemblablement, le vrai Soleil central.*"

(19) p. 39. — See the article *On Gravitation*, Penny Cyclopædia, from the pen of the Astronomer Royal.

(20) p. 41. — Delambre dates the commencement of modern astronomical observation in its most perfect form from Maskelyne, who was the first who gave what is now called a standard catalogue (A.D. 1790) of stars ; that is, a number of stars observed with such frequency and accuracy, that their places serve as standard points of the heavens. His suggestion of the *Nautical Almanack*, and his superintendence of it to the end of his life, from its first publication in 1767, are mentioned in the *Almanack* (vol. i. p. 364) ; his *Schehallion Experiment on Attraction*, in vol. iii. p. 69 ; and the character of his *Greenwich Observations in Greenwich Observatory*, in vol. ii. p. 442.

(21) p. 41. — *Experiments to determine the density of the Earth.* By Henry Cavendish, Esq., F.R.S. and F.A.S. — Philosophical Transactions, 1798.

(22) p. 42. — Adams. *An Explanation of the observed irregularities in the motion of Uranus, on the hypothesis of disturbance caused by a more distant Planet.* — Appendix to Nautical Almanack for 1851.

(23) p. 43. — Le Verrier : *Premier Mémoire sur la théorie d'Uranus*, Comptes Rendus, vol. xxi. ; *Sur la planète qui produit les anomalies observées dans le mouvement d'Uranus.* — *Ib.* vol. xxiii.

(24) p. 43. — The experiment alluded to is one of a series by M. Plateau, who thus describes his arrangement of the fluid : "We begin by making a mixture of alcohol and distilled water, containing a certain excess of alcohol, so that, when submitted to the trial of the

test tube, it lets the small sphere of oil fall to the bottom rather rapidly. When this point is obtained, the whole is thrown upon filters, care being taken to cover the funnels containing these last with plates of glass; this precaution is necessary in order to prevent, as much as possible, the evaporation of the alcohol. The alcoholic liquor passes the first through the filters, ordinarily carrying with it a certain number of very minute spherules of oil. When the greater part has thus passed, the spherules become more numerous; what still remains in the first filters, namely, the oil and a residue of alcoholic liquor, is then thrown into a single filter placed on a new flask. This last filtration takes place much more slowly than the first, on account of the viscosity of the oil; it is considerably accelerated by renewing the filter once or twice during the operation. If the funnel has been covered with sufficient care, the oil will collect into a single mass at the bottom of the flask under a layer of alcoholic liquor." — *On the phenomena presented by a free Liquid Mass withdrawn from the action of Gravity.* By Professor Plateau, of the University of Ghent. Translated from the *Memoirs of the Royal Academy of Brussels*, vol. xvi.; in the *Scientific Memoirs*, vol. iv. part 13.

([25]) p. 47. — "The divisibility of matter is great beyond the power of imagination, but we have no reason for asserting that it is infinite; for the demonstrations which have sometimes been adduced in favour of this opinion are obviously applicable to space only. The infinite divisibility of space seems to be essential to the conception that we have of its nature, and it may be strictly demonstrated that it is mathematically possible to draw an infinite number of circles between any given circle and its tangent, none of which shall touch either of them except at the general point of contact; and that a ship following always the same oblique course with respect to the meridian, — for example, sailing north-eastwards, — would continue perpetually to approach the pole without ever completely reaching it. But when we inquire into the truth of the old maxim of the schools, that all matter is infinitely divisible, we are by no means able to decide so positively. Newton observes that it is doubtful whether any human means may be sufficient to separate the particles of matter beyond a certain limit; and it is not impossible that there may be some constitution of atoms, or single corpuscles, on which their properties, as matter, depend, and which would be destroyed if the units were further divided; but it appears to be more probable that there are no such atoms, and even if there are, it is almost certain that matter is never thus annihilated in the common course of nature." — *The Essential*

Properties of Matter: Young's *Natural Philosophy;* ed. by Rev. P. Kelland.

([26]) p. 47. — "Two very different hypotheses have been formed to explain the nature of matter, or the mode of its formation; the one known as the *atomic* theory, the other, the *dynamic.* The founder of the former and earlier was Leucippus: he considered the basis of all bodies to be extremely fine particles, differing in form and nature, which he supposed to be dispersed throughout space, and to which his follower Epicurus first gave the name of atoms. To these atoms he attributed a rectilinear motion, in consequence of which such as are homogeneous united, whilst the lighter were dispersed through space. The author of the second hypothesis was the famous Kant. He imagined all matter existed, or was originated, by two antagonist and mutually counteracting principles, which he called attraction and repulsion, all the predicates of which he referred to motion. Most modern philosophers, and foremost amongst them Ampère and Poisson, have adopted an hypothesis combining the features of both the preceding. They regard the atoms as data, deriving their origin from the Deity as the first cause, and consider their innate attractive and repulsive force as a necessary condition to their combination in bodies. The main features of this hypothesis are borrowed from Aristotle, inasmuch as he supposed the basis of all bodies to be the four elements known to the ancients, the particles of which, endued with certain powers, constituted bodies. According to Ampère, all bodies consist of equal particles, and they again of molecules that, up to a certain distance, attract each other. Their distance from each other he supposed to be regulated by the intensity of the attractive and repulsive forces, the latter of which preponderates." — Peschel's *Elements of Physics;* translated by E. West. 1845.

([27]) p. 48. — This was first proved by the researches of Dr. Dalton; the subject will be again alluded to under the consideration of atomic volumes.

([28]) p. 49. — These peculiar phenomena may be studied advantageously in the works of most of the eminent European chemists. In our own language, the reader is referred to Dr. Thompson's *Outline of the Sciences of Heat and Electricity,* 2nd Edition; Brande's *Manual of Chemistry* — Art. *Specific Heat;* Graham's *Elements of Chemistry;* and Daniel's *Introduction to the Study of Natural Philosophy.*

([29]) p. 49. — The conversion of the diamond into graphite and coke was first effected by the agency of the galvanic arc of flame, by M. Jaquelini, and communicated to the Academy of Sciences in 1847, in a Memoir entitled, *D l'action calorifique de la pile de Bunsen, du chalumeau à gaz oxygène et hydrogène sur le carbon pur, artificiel et naturel.* See *Comptes Rendus*, 1847, vol. xxiv. p. 1050; also *Report of the British Association*, for 1847, (*Transactions of Sections*,) p. 50.

([30]) p. 50.—"In the annual report on the progress of chemistry, presented to the Royal Academy of Stockholm, in March 1840, I have proposed to designate by the term *allotropic state*, that dissimilar condition which is observed in certain elements, and long known examples of which are found in the different forms of carbon, as graphite and diamond.

"Although these dissimilar conditions, which I have here called allotropic, have long since attracted attention in one or two elements, still they have been regarded as exceptions to the general rule. It is at present my object to show that they are not so rare; that it is probably rather a general property of the elements to appear in different allotropic conditions; and that, although we have hitherto been unable to obtain several of the elements when uncombined in their allotropic states, still their compounds indicate the same with tolerable distinctness."—*Berzelius on the Allotropy of the Elementary Bodies, &c.:* Poggendorff's Annalen, 1844. Scientific Memoirs, vol. iv. p. 240.

([31]) p. 50.—"Copper, when reduced by hydrogen at a heat below that of redness, on exposure to air soon becomes converted throughout its mass into protoxide; and when it is triturated for some time with an equivalent quantity of sulphur, it combines with it according to Böttcher's experiment, producing flame, and forming sulphuret of copper. If, however, the copper be reduced by hydrogen at a red heat, still considerably below the temperature at which it softens and begins to melt, it remains for years unchanged by exposure to air, and cannot be made to combine with sulphur without the application of heat. Iron, cobalt, and nickel, when reduced by hydrogen below a red heat, inflame after they have cooled, if exposed to the air; and if they are immediately placed in water to avoid their taking fire, they inflame when they are again removed, and become nearly dry. If we compare this behaviour with that of iron reduced by heat, and with iron in that state in which it forms the conductor of a galvanic current without becoming oxidized, it would appear that these peculiarities depended upon something more than a difference of mechanical condition."—*Berzelius on the Allotropy of Elementary Bodies.* See *On the Isomeric Conditions of the Peroxide of Tin:* by Prof. H. Rose.—Chemical Gazette, Oct. 1848.

([32]) p. 51.—On this curious subject, and its history, see Bergman's *Dissert. de Phlog. quantitate in Metallis*, 1764. Kirwan, *On the Attractive Powers of Mineral Acids:* Philosophical Transactions. Kier's *Experiments and Observations on the Dissolution of Metals in Acids:* Phil. Trans. 1790.

From these valuable papers it will be seen that the peculiar states of iron had already attracted attention, particularly those "inactive conditions" noticed in a "*Note sur la Manière d'agir de l'Acide nitrique sur le Fer, par J. F. W. Herschel*," Aug. 1833; and previously indicated by M. H. Braconnot, *Sur quelques Propriétés de l'Acide nitrique*, Annales de Chimie, vol. lii. p. 54. Reference should also be made to the Memoirs of Sir John Herschel, *On the Action of the Rays of the Solar Spectrum on Vegetable Colours, &c.:* Phil. Trans. vol. cxxxiii. p. 221; and *On the Separation of Iron from other Metals:* Phil. Trans. vol. cxi. p. 293; and several papers by Schönbein, in the Philosophical Magazine, from 1837.

([33]) p. 51.—Faraday, in his memoir *On new Magnetic Actions, and on the Magnetic Condition of all Matter*, says: "By the exertion of this new condition of force, the body moved may pass either along the magnetic lines or across them, and it may move along or across them in either or any direction, so that two portions of matter, simultaneously subject to this power, may be made to approach each other as if they were mutually attracted, or recede as if mutually repelled. All the phenomena resolve themselves into this, that a portion of such matter, when under magnetic action, tends to move from stronger to weaker places or points of force. When the substance is surrounded by lines of magnetic force of equal power on all sides, it does not tend to move, and is then in marked contradistinction with a linear current of electricity under the same circumstances." — Phil. Trans. for 1846, vol. cxxxvii.

([34]) p. 52. — *New Experiments and Observations on Electricity, made at Philadelphia, in America.* — Addressed to Mr. Collinson, from 1747 to 1754. By Benjamin Franklin. Of these Priestley remarks: "It is not easy to say whether we are most pleased with the simplicity and perspicuity with which the author proposes every hypothesis of his own, or the noble frankness with which he relates his mistakes, when they were corrected by subsequent experiments."

([35]) p. 52. — "The atomic philosophy of Epicurus, in its mere physical contemplation, allows of nothing but matter and space, which are equally infinite and unbounded, which have equally existed from all eternity, and from different combinations of which every visible form is created. These elementary principles have no common property with each other; for whatever matter is, that space is the reverse of; and whatever space is, matter is the contrary to. The actual solid parts of all bodies, therefore, are matter, their actual pores space, and the parts which are not altogether solid, but an intermixture of solidity and pore, are space and matter combined.

"These infinite groups of atoms, flying through all time and space in different directions and under different laws, have interchangeably tried and exhibited every possible mode of rencounter: sometimes repelled from each other by concussion, and sometimes adhering to each other from their own jagged or pointed construction, or from the casual interstices which two or more connected atoms must produce, and which may be just adapted to those of other figures, as globular, oval, or square. Hence the origin of compound and visible bodies; hence the origin of large masses of matter; hence, eventually, the origin of the world itself."— Dr. Good's *Book of Nature.*

([36]) p. 53. — *A Course of Lectures on Natural Philosophy and the Mechanical Arts.* By Thomas Young, M.D. Lecture 49, *On the essential properties of Matter.*

([37]) p. 57. — "Gay-Lussac first made the remark, that a crystal of potash alum, transferred to a solution of ammonia alum, continued to increase without its form being modified, and might thus be covered with alternate layers of the two alums, preserving its regularity and proper crystalline figure. M. Beudant afterwards observed that other bodies, such as the sulphates of iron and copper, might present themselves in crystals of the same form and angles, although the form was not a simple one, like that of alum. But M. Mitscherlich first recognized this correspondence in a sufficient number of cases to prove that it was a general consequence of similarity of composition in different bodies." — Graham's *Elements of Chemistry* (1842), p. 136.

The following remarks are from a paper by Dr. Hermann Kopp, *On the Atomic Volume and Crystalline Condition of Bodies*, &c, published in the Philosophical Magazine for 1841: "The doctrine of isomorphism shows us that there are many bodies which possess an analogous constitution, and the same crystalline form. Our idea of the volume (or, in other words, of the crystalline form) of these bodies must therefore be the same. From this it follows that their specific weight is dependent upon our idea of mass, that is, of atomic weight, whilst our idea of specific weight is connected with mass contained in the same volume. From these considerations the following law may be deduced: *The specific weight of isomorphous bodies is proportional to their atomic weight, or isomorphous bodies possess the same atomic volume.*" — p. 255. A translation appears in the works of the Cavendish Society, from Dr. Otto's Chemistry, *On Isomorphism*, which may be advantageously consulted. See also a paper by M. Rose, translated from the *Proceedings of the Royal Berlin Academy* for the *Chemical Gazette*, Oct. 1848, entitled, *On the Isomeric Conditions of the Peroxide of Tin.*

(38) p. 57. — *A System of Mineralogy, comprising the most recent discoveries:* by James D. Dana, A.M., New York, 1844.

(39) p. 57. — Crystallogeny, or the formation of crystals, is the term employed by Dana, in his admirable work quoted above: whose remarks on *Theoretical Crystallogeny*, p. 71, are well worthy of all attention.

(40) p. 58. — *On the Magnetic relations of the Positive and Negative Optic Axes of Crystals:* by Professor Plücker, of Bonn. — Philosophical Magazine, No. 231 (3rd Series), p. 450. *Experimental Researches on Electricity; On the crystalline polarity of Bismuth and other bodies, and on its relation to the magnetic form of force:* by Michael Faraday, Esq., F.R.S. — Transactions of the Royal Society for 1848.

(41) p. 58. — In the *Memoirs of the Geological Survey of the United Kingdom, and of the Museum of Economic Geology*, vol. i. 1846, will be found a paper, by the author of this volume, *On the Influences of Magnetism on Crystallization, and other Conditions of Matter*, in which the subject is examined with much care. See also *Magnétisme polaire d'une montagne de Chlorite schisteuse et de Serpentine:* Annales de Chimie, vol. xxv. p. 327; *Influence du Magnétisme sur les actions chimiques*, by l'Abbé Rendus; and also a notice of the experiments of Ritter and Hansteen, "Analysées par M. Œrsted;" also *Effets du Magnétisme terrestre sur la précipitation de l'Argént, observées par M. Muschman:* Annales de Chimie, vol. xxxviii. p. 196–201.

(42) p. 60. — The transparent varieties of sulphate of lime are distinguished by the name *Selenite;* and the fine massive varieties are called *Alabaster*. Gypsum often forms very extensive beds in secondary countries, and is also found in tertiary deposits; occasionally, in primitive rocks; it is also a product of volcanoes. The finest foreign specimens are found in the salt mines of Bex, in Switzerland; at Hall, in the Tyrol; in the Sulphur-mines of Sicily; and

in the gypsum formation near Ocana in Spain. In England, the clay of Shotover Hill, near Oxford, yields the largest crystals. See Dana's *Mineralogy*, second edition, p. 241.

([43]) p. 54. — The following table of the rays penetrating coloured glass has been given by Melloni, in his memoir *On the Free Transmission of Radiant Heat through Different Bodies.*

Deep violet	53	Deep yellow	40
Yellowish red (flaked)	53	Bright yellow	34
Purple red (flaked)	51	Golden yellow	33
Vivid red	47	Deep blue	33
Pale violet	45	Apple green	26
Orange red	44	Mineral green	23
Clear blue	42	Very deep blue	19

Translated in the Scientific Memoirs, vol. i. p. 30.

([44]) p. 64. — "The physical characters of this species of glass, which acts so differently from the other species of coloured glass in all the phenomena of calorific absorption, are, 1st, its intercepting almost totally the rays which pass through alum; 2nd, its entirely absorbing the red rays of the solar spectrum. I have already stated that their colouration is produced almost entirely by the oxide of copper.

"Thus, the colouring matters of the coloured glasses, while they so powerfully affect the relations of quantity which the different rays of ordinary light bear to each other, exercise no elective action on the concomitant calorific rays. This curious phenomenon is the more remarkable as the colouring matters absorb almost always a very considerable portion of the heat *naturally transmitted by the glass.* The following are, in fact, the calorific transmissions of the seven coloured glasses referred to; the transmission of the common glass being represented by 100; red glass, 82.5; orange, 72.5; yellow, 55; bluish-green, 57.5; blue, 52.5; indigo, 30; violet, 85. The quantity of heat absorbed through the action of the colouring substances is, therefore, 17.5 in the red glass, 27.5 in the orange, 45 in the yellow, 42.5 in the green, 47.5 in the blue, 70 in the indigo, and 15 in the violet. Now, as these absorptions extinguish a proportional part of each of the rays which constitute the calorific stream transmitted by common glass, they may be compared, as we said before, with the absorbent action exercised on light by matters more or less deeply brown or dark, when they are immersed in water, or some other colourless liquid which dissolves, but does not affect them chemically." — *Annales de Chimie et de Physique*, tom. xl. p. 382.

Guided by these principles, the author selected the glass employed in glazing the Royal Palm-House, at Kew Botanical Gardens, where it was desired to obstruct the passage of those rays which have a particular scorching influence. Of this glass a description was given at the meeting of the British Association at Oxford, and appears in the Transactions for that year; and a paper more in detail will be found in the forthcoming Transactions of the Society of Arts.

([45]) p. 65. — In the *Philosophical Transactions*, vol. xc., the following papers, by Sir William Herschel, may be consulted: —

Investigation of the powers of the prismatic colours to heat and illuminate objects; with remarks that prove the different refrangibility of radiant heat. To which is added, an inquiry into the method of viewing the sun advantageously, with telescopes of large apertures and high magnifying powers, p. 255. *Experiments on the refrangibility of the invisible rays of the sun*, p. 284. *Experiments on the solar and on the terrestrial rays that occasion heat; with a comparative view of the laws to which light and heat, or rather the rays which occasion them, are subject, in order to determine whether they are the same or different*, pp. 293, 437.

In connection with this inquiry, Sir William Herschel remarks, that since a *red glass* stops no less than 692 out of 1,000 such rays as are of the refrangibility of red light, we have a direct and simple proof, in the case of the red glass, that the rays of light are transmitted, while those of heat are stopped, and that thus they have nothing in common but a certain equal degree of refrangibility, which by the power of the glass must occasion them to be thrown together into the place which is pointed out to us by the visibility of the rays of light.

On the same subject, a Memoir, by Sir Henry Englefield, in the Journal of the Royal Institution for 1802, p. 202, may be consulted; and *Researches on Light*, by the Author.

([46]) p. 65. — Dr. Draper, *On the production of light by heat*, in the Phil. Mag. for 1847.

Sir Isaac Newton fixed the temperature at which bodies become self-luminous at 635°; Sir Humphrey Davy at 812°; Mr. Wedgewood at 947°; and Mr. Daniell at 980°; whilst Dr. Draper, from his experiments, gives 977°; and Dr. Robinson 865°.

In a review of the above paper by Melloni, entitled, *Researches on the Radiations of Incandescent Bodies, and on the Elementary Colours of the Solar Spectrum*, translated from Silliman's Journal for August, 1847, he remarks: —

"I say that they conduct, as do others heretofore known on light and radiant heat, to a perfect analogy between the general laws which govern these two great agents of nature. I will add, that I regard the theory of their identity as the only one admissible by the rules of philosophy; and that I consider myself obliged to adopt it, until it shall have been proved to me that there is a necessity of having recourse to two different principles, for the explanation of a series of phenomena which at present appear to belong to a solitary agent."

Reference should also be made to a paper by Dr. Robinson, *On the effects of Heat in lessening the Affinities of the Elements of Water*, in the Transactions of the Royal Irish Academy, 1848, where he says that "when a platinum wire is traversed by a current gradually increased till it produces ignition, the first gleam that appears is not red, but of a colour which, when I first saw it, I compared to the 'lavender ray' discovered by Sir John Herschel beyond the violet, though I was surprised at seeing the tint of that most refrangible ray preceding the ray which is least so. It is quite conspicuous at about 865°; and as the mode in which it makes its appearance presents nothing abrupt or discontinuous, it seems likely that it is merely a transition from invisible rays excited at a lower temperature to ordinary light."—p. 310.

([47]) p. 66.—In the *Bakerian Lecture* for 1842, *On the transparency of the Atmosphere, and the law of extinction of the solar rays in passing through it*, by James D. Forbes, Esq., F. R. S., &c., will be found a most complete investigation of this subject.

The experiments were, for the most part, made in Switzerland with Sir John Herschel's actinometer, and they prove most satisfactorily,—"That the absorption of the solar rays by the strata of air to which we have immediate access, is considerable in amount for even moderate thicknesses."

([48]) p. 66.—After referring to several curious and instructive experiments, in which peculiar chemical changes are produced under the influence of the solar rays by their Heat, Sir John Herschel says:—

"These rays are distinguished from those of Light by being invisible; they are also distinguished from the purely calorific rays beyond the spectrum, by their possessing properties *(of a peculiar character, referred to in former papers)* either exclusively of the calorific rays, or in a much higher degree. They may perhaps not improperly be regarded as bearing the same relation to the calorific spectrum which the photographic rays do to the luminous ones. If the restriction to these rays of the term *thermic*, as distinct from *calorific*, be not (as I think, in fact, it is not) a sufficient distinction, I would propose the term *parathermic rays* to designate them. These are the rays which I conceive to be active in producing those singular molecular affections which determine the precipitation of vapours in the experiments of Messrs. Draper, Moser, and Hunt, and which will probably lead to important discoveries as to the intimate nature of those forces resident on the surfaces of bodies, to which M. Dutrochet has given the name of *epipolic* forces."— *On certain improvements in Photographic processes, described in a former communication* (Phil. Trans. vol. cxxxiii.); and *On the Parathermic Rays of the Solar Spectrum:* Phil. Trans. vol. cxxxiv.

The experiments of Mrs. Somerville, *On the action of the Rays of the Spectrum on Vegetable Juices* (Phil. Transactions, vol. cxxxvii.), appear to connect themselves with this particular class of rays in a curious manner.

(49) p. 67.—Experiments on the influence of heat on differently-coloured bodies were first made by Dr. Hooke; and it was not until long after that Franklin made his ingenious experiments. Davy exposed to sunshine six equal pieces of copper, painted white, yellow, red, green, blue, and black, in such a manner that one side only was illuminated. To the dark side he attached a bit of cerate, ascertained by experiment to melt at 70°. The cerate attached to the black became fluid first, the blue next, then the green and red, and lastly the yellow and white.—Beddoes's *Contributions to Physical Knowledge*, and collected works of Sir Humphrey Davy, vol. ii. p. 27.

(50) p. 68.—By reference to the Treatise on Heat, in the *Encyclopædia Metropolitana*, numerous suggestive experiments will be found, all bearing on this subject. Peschel's *Elements of Physics* may also be consulted with advantage. The fact is, however, simply proved by placing the bulbs of delicate thermometers, so as to be completely involved in the petals of flowers exposed to sunshine, shading the upper portion of the stem of the instrument.

(51) p. 68.—Moser, *On Vision, and on the action of Light on Bodies:* and also *On Latent Light:* Scientific Memoirs, vol. iii. Draper, *On certain Spectral Appearances, and on the discovery of Latent Light*; Phil. Mag. Nov. 1842.

(52) p. 68.—A particular examination of this curious question will be found in the Author's report *On the influence of the Solar Rays on the Growth of Plants:* Reports of the British Association for 1847.

(53) p. 71.—Ammianus Marcellinus ascribes the longevity and robust health of mountaineers to their exposure to the dews of night.

Dew was employed by the alchemists in their experiments on the solution of gold. The ladies of old collected the "celestial wash," which they imagined had the virtue of preserving their fine forms, by exposing heaps of wool to the influences of night radiation. It was supposed that the lean features of the grasshopper arose from that insect feeding entirely on dew, "Dumque thymo pascentur apes, dum rore cicadæ,' Virgil, Eclog.

See some curious remarks by Boyle, *On the power of Dew in working on Solid Bodies:* Works of the Honorable R. Boyle, vol. v. p. 121, 1744.

([54]) p. 72.—See the *Researches on Heat*, by Professor James Forbes, in the Transactions of the Royal Society of Edinburgh; also Melloni's papers on the same subject in the *Annales de Chimie*, several of which have been translated into the *Scientific Memoirs*, edited by Mr. Richard Taylor.

([55]) p. 72.—The phenomena of dew have constantly engaged the attention of man. Aristotle, in his book *De Mundo*, puts forth some just notions on its nature. An opinion has almost always prevailed that dew falls. Gersten appears to have been the first who opposed this motion. He was followed by Muschenbroeck, and then by Du Fay. The researches of Leslie were of a far more exact character. Dr. Wilson, in the Transactions of the Royal Society of Edinburgh, 1st vol., published a *Memoir on Hoar Frost*, of much interest; but the questions involved remained unsettled until the researches of Dr. Wells, which were published in his *Essay on Dew*, in 1814.

([56]) p. 72.—By far the most complete set of experiments on the radiation of heat from the surface at night, which have been published since Dr. Wells's memoir *On Dew*, are those of Mr. Glaisher of the Royal Observatory at Greenwich. Instruments of the most perfect kind were employed, and the observations made with sedulous care. The results will be found in a memoir *On the Amount of the Radiation of Heat, at night, from the Earth and from various bodies placed on or near the Surface of the Earth*, by James Glaisher, Esq.: Philosophical Trans. for 1847, part 2.

([57]) p. 73.—Dr. Wells noticed the practical fact that very light shades protected delicate plants from frost, by preventing radiation. Mr. Goldsworthy Gurney has made a series of interesting experiments, and he imagines that by shading grass-lands with boughs of trees, or any light litter, a more abundant crop is produced. The subject has been discussed in the journals of the Royal Agricultural

Society. May not the apparent increase be due entirely to the succulent condition in which a plant always grows in the shade?

([58]) p. 74.—This paper of Melloni's will be found in the *Bibliothèque Universelle de Genève* for 1843. The conclusions are highly ingenious, but they rest entirely on the analogy supposed to be discovered between the relations of heat, like light, to the coloured rays of the spectrum. This, it must be remembered, is not the case, since even Sir William Herschel showed that red light might exist with only a minimum of calorific power, notwithstanding the fact that the maximum heat-ray of the spectrum coincides with the red rays.

([59]) p. 76.—Dr. Robinson, of Armagh, in his memoir *On the Effects of Heat in lessening the Affinities of the Elements of Water.*—Transactions of the Royal Irish Academy, vol. xxi. part 2.

([60]) p. 77.—On this subject consult Robert Were Fox, *On the Temperature of the Mines of Cornwall.*—Cornwall Geological Transactions, vol. ii.; W. J. Henwood, on the same subject, *Ib.* vol. v.; Reports of the British Association, 1840, p. 315; Edinburgh New Philosophical Journal, vol. xxiv. p. 140.

([61]) p. 78.—*On the causes of the temperature of Hot and Thermal Springs; and on the bearings of this subject as connected with the general question regarding the internal temperature of the Earth:* by Professor Gustav Bischoff, of Bonn.—Edinburgh New Philosophical Journal, vol. xx. p. 376; vol. xxiii. p. 330. Some interesting information on the temperature of the ground will be found in Erman's *Travels in Siberia*, translated by W. D. Cooley, vol. i. p. 339; vol. ii. p. 366. *Sur la Profondeur à laquelle se trouve la couche de Température invariable entre les Tropiques*, by Boussingault: Annales de Chimie et de Physique, 1833, p. 225. Reference may also be made to Humboldt's *Cosmos*, Sabine's translation, vol. i. p. 165, &c.; and to the excellent article on *Meteorology*, by George Harvey, in the Encyclopædia Metropolitana. These chthonisothermal lines are, however, materially influenced by the form of the surface, curving upwards under a mountain, and downwards under extensive valleys.

([62]) p. 78.—These results are obtained from the valuable observations of Robert Were Fox, Esq., made with great care by that gentleman in several of the Cornish mines: *Report on some Observations on Subterranean Temperature.*—British Association Reports, vol. ix. p. 309; Philosophical Magazine, 1837, vol. ii. p. 520.

([63]) p. 80.—From his experiments, the following conclusions were arrived at by M. Delaroche:—

1. Invisible radiant heat may, in some circumstances, pass directly through glass.

2. The quantity of radiant heat which passes directly through glass is so much greater, relative to the whole heat emitted in the same direction, as the temperature of the source of heat is more elevated.

3. The calorific rays which have already passed through a screen of glass, experience, in passing through a second glass screen of a similar nature, a much smaller diminution of their intensity than they did in passing through the first screen.

4. The rays emitted by a hot body differ from each other in their faculty to pass through glass.

5. A thick glass, though as much or more permeable to light than a thin glass of worse quality, allows a much smaller quantity of radiant heat to pass. The difference is so much the less as the temperature of the radiating source is more elevated.

6. The quantity of heat which a hot body yields in a given time, by radiation to a cold body situate at a distance, increases, *cæteris paribus*, in a greater ratio than the excess of temperature of the first body above the second.—Journal de Physique, vol. lxxv.

([64]) p. 80.—Sir David Brewster differs from the conclusions arrived at by Delaroche. He thus explains his views: "The inability of radiant heat to pass through glass, may be considered as a consequence of its refusing to yield to the refractive force; for we can scarcely conceive a particle of radiant matter freely permeating a solid body, without suffering some change in its velocity and direction. The ingenious experiments of M. Prévost, of Geneva, and the more recent ones of M. Delaroche, have been considered as establishing the permeability of glass to radiant heat. M. Prévost, employed movable screens of glass, and renewed them continually, in order that the result which he obtained might not be ascribed to the heating of the screen; but such is the rapidity with which heat is propagated through a thin plate of glass, that it is extremely difficult, if not impossible, to observe the state of the thermometer before it has been affected by the secondary radiation from the screen. The method employed by M. Delaroche, of observing the difference of effect, when a blackened glass screen and a transparent one were made successively to intercept the radiant heat, is liable to an obvious error. The radiant heat would find a quicker passage through the transparent screen; and, therefore, the difference of effect was not due to the transmitted heat, but to the heat radiated from the anterior surface. The truth contained in M. Delaroche's fifth proposition is almost a demonstration of the fallacy of all those that precede it. He found that 'a thick plate of glass, though as much or more permeable to light than a thin glass of worse quality, allowed a much smaller quantity of radiant heat to pass.' If he had employed very thick plates of the purest flint glass, or thick masses of fluid that have the power of transmitting light copiously, he would have found that not a single particle of heat was capable of passing directly through transparent media."—Sir D. Brewster, *On new properties of heat as exhibited in its propagation along plates of glass.* Philosophical Transactions, vol. cvi. p. 107.

(65) p. 80.—*Proposal of a New Nomenclature for the Science of Calorific Radiations*, by M. Melloni. Bibliothèque Universelle de Genève, No. 70. Scientific Memoirs, vol. iii. part 12. Many of the terms, as *Diathermasy*, or transparency for heat; *Adiathermasy*, opacity for heat; *Thermochroic*, coloured for heat, and others, are valuable suggestions of forms of expression which are required in dealing with these physical phenomena.

(66) p. 81.—For a careful examination of the several theories of heat, consult Dr. Young's Course of Lectures on Natural Philosophy, &c., Lecture 52, *On the Measures and the Nature of Heat;* also Powell's very excellent *Reports on Radiant Heat*—Reports of the British Association, 1832, 1840. The transcendental view which the immaterial theory leads to, cannot be better exemplified than by the following quotation from that inexplicable dream of a talented man, *Elements of Physiophilosophy*, by Lorenz Oken, M. D. (translated for the Ray Society, by Alfred Tulk):—

"Heat is not matter itself any more than light is; but it is only the act of motion in the primary matter. In heat, as well as in light, there certainly resides a material substratum; yet, this substratum does not give out heat and light; but the *motion* only of the substratum gives out heat, and the *tension* only of the substratum light. There is no body of heat; nitrogen is the body of heat, just as oxygen may be called the body of fire. Heat is real space; into it all forms have been resolved, as all materiality has been resolved into gravity, and all activity, all polarity into light. Heat is the universal form, consequently the want of form."

(67) p. 81. — Mémoires de la Société Physiqué, &c., Genève, tom. ii. part 2.

(68) p. 82. — This curious phenomenon was first observed by Mr. Trevelyan, whose *Notice regarding some Experiments on the Vibration of Heated Metals* will be found in the Transactions of the Royal Society of Edinburgh, vol. xii. 1837. In a Memoir in the same volume, entitled *Experimental Researches, regarding certain vibrations which take place between metallic masses having different temperatures*, Professor Forbes draws the following conclusions: —

1. "The vibrations never take place between substances of the same nature.

2. "Both substances must be metallic.

3. "The vibrations take place with an intensity proportional (within certain limits) to the difference of the conducting powers of the metals for heat or electricity; the metal having the least conducting power being necessarily the coldest.

4. "The time of contact of two points of the metals must be longer than that of the intermediate portions.

5. "The impulse is received by a distinct and separate process at each contact of the bar and block, and in no case is the metallic connection of the bearing points in the bar, or those of the block, in any way essential.

6. "The intensity of the vibration is (under certain exceptions) proportional to the difference of temperature of the metals." — Transactions of the Royal Society of Edinburgh, vol. xii.

(69) p. 82. — The Bakerian Lecture. *On certain Phenomena of Voltaic Ignition, and the Decomposition of Water into its Constituent Gases by Heat*: by W. R. Grove, Esq. — Philosophical Transactions, 1847. Part 1.

(70) p. 83. — Davy's *Researches on Flame*. Works, vol. vi. — Philosophical Transactions for 1817.

(71) p. 83. — *On the Effect of Heat in lessening the affinities of the Elements of Water*: by the Rev. Thomas Romney Robinson, D.D. — Transactions of the Royal Irish Academy, vol. xxi. part 2.

(72) p. 83. — *An Inquiry concerning the Chemical properties that have been attributed to Light*: by Benjamin, Count of Rumford. — Philosophical Transactions, vol. lxxxviii. p. 449. — The results obtained by Count Rumford were probably due to the non-luminous heat-rays — parathermic rays — which are known to be given off by boiling water.

([73]) p. 83. — For Dr. Draper's paper, see Philosophical Magazine for May, 1847, vol. xxx. 3rd series.

([74]) p. 84. — *On the action of the rays of the Solar Spectrum on Vegetable Colours:* by Sir J. F. W. Herschel, Bart.

The proof of the continuation of the visible prismatic spectrum beyond the extreme violet may be witnessed in the following manner: "Paper stained with tincture of turmeric is of a yellow colour; and, in consequence, the spectrum thrown on it, if exposed in open daylight, is considerably affected in its apparent colours, the blue portion appearing violet, and the violet very pale and faint; but beyond the region occupied by the violet rays, is distinctly to be seen a faint prolongation of the spectrum, terminated laterally, like the rest of it, by straight and sharp outlines, and which, in this case, affects the eye with the sensation of a pale yellow colour." — Philosophical Transactions, p. 133.

([75]) p. 86. — The most complete exposition of the theory that animal heat is derived from chemical action only, will be found in *Animal Chemistry, or Chemistry in its applications to Physiology and Pathology*, by Justus Liebeg: translated by Dr. Gregory. The conclusions arrived at by the author, notwithstanding his high — and deservedly high — position in chemical science, must, however, be received with great caution, many of them being founded on most incorrect premises, and his generalizations are of the most hasty and imperfect character. At page 22, the following passage occurs: "If we were to go naked, like certain savage tribes, or if in hunting or fishing we were exposed to the same degree of cold as the Samoiedes, we should be able, with ease, to consume ten pounds of flesh, and, perhaps, a dozen of tallow candles into the bargain daily, as warmly-clad travellers have related with astonishment of these people. We should then also be able to take the same quantity of brandy or train-oil without bad effects, because the carbon and hydrogen of these substances would only suffice to keep up the equilibrium between the external temperature and that of our bodies."

A brief examination will exhibit the error of this. The analysis of Beef, by Dr. Lyon Playfair, is as follows: —

Carbon	51.83
Hydrogen	7.57
Nitrogen	15.01
Oxygen	21.37
Ashes	4.23

And the following has been given by Chevreul as the composition of mutton tallow:—

Carbon	96
Hydrogen	16
Nitrogen	16
Oxygen	48

About three times the quantity of oxygen to the carbon eaten, is required to convert it into carbonic acid; hence, the Samoiede, eating more highly carbonized matter, must inspire 288 ozs. of oxygen daily, or nearly eight times as much as the "ordinary adult." By the lungs he must take into the body 2,304 cubic feet of air besides what will be absorbed by the skin. His respirations must be so much quickened, that at the lowest possible calculation he must have 500 pulsations a minute. Under such conditions it is quite clear man could not exist. There is no disputing the fact of the enormous appetites of these people; but all the food is not removed from the system as carbonic acid gas.

([76]) p. 87.—An interesting paper by Dr. Davy, *On the Temperature of Man*, will be found in the Philosophical Transactions, vol. cxxxvi. p. 319.—Sir Humphrey Davy, in his *Consolations in Travel, or the Last Days of a Philosopher*, in his fourth dialogue, *The Proteus*, has several ingenious speculations on this subject.

([77]) p. 87.—*Exposé de quelques résultats obtenus par l'action combinée de la chaleur et de la compression sur certains liquides, tels que l'eau, l'alcool, l'éther sulfurique, et l'essence de pétrole rectifiée:* par M. le Baron Cagniard de la Tour.

The three following conclusions are arrived at:—

1. Que l'alcool à 36 degrés, l'essence de pétrole rectifiée à 42 degrés, ét l'éther sulfurique soumis à l'action de la chaleur et de la compression, sont susceptibles de se réduire complètement en vapeur sous un volume un peu plus que double de celui de chaque liquide.

2. Qu'une augmentation de pression, occasionnée par la présence de l'air dans plusieurs des expériences qui viennent d'étre citées, n'a point apporté d'obstacle à l'évaporation du liquide dans le même espace; qu'elle a seulement rendu sa dilatation plus calme et plus facile à suivre jusqu'au moment où le liquide semple s'évanouir tout-à-coup.

3. Que l'eau, quoique susceptible sans doute d'être réduite en vapeur très-comprimée, n'a pu être soumise à des expériences complètes, faute de moyens suffisans pour assurer l'exacte ferme-

ture de la marmite de compression, non plus que dans les tubes de verre dont elle altère la transparence en s'emparant de l'alcali qui entre dans leur composition.—Annales de Chimie, vol. xxi.

([78]) p. 88.—*Sur les phénomènes que présentent les corps projetés sur des surfaces chaudes:* par M. Boutigny (d'Evreux).—Annales de Chimie et de Physique, vol. xi. p. 16. *Congélation du mercure en trois secondes, en vertu de l'état sphéroïdal dans un creuset incandescent:* par M. Faraday.—Ibid., vol. xix. p. 383.

Spheroidal Condition of Bodies (Extrait d'une Note de M. Boutigny d'Evreux).

"Au nombre des propriétés des corps à l'état sphéroïdal, il en est cinq qui me paraissent caractéristiques et fondamentales, et c'est sur ces cinq propriétés que je base la définition que je soumets aujourd'hui au jugement de l'Académie. Ces cinq propriétés sont:—

"1. La forme arrondie que prend la matière sur une surface chauffée à une certaine température.

"2. Le fait de la distance permanente qui existe entre le corps à l'etat sphéroïdal et le corps sphéroïdalisant.

"3. La propriété de réfléchir le calorique rayonnant.

"4. La suspension d l'action chimique.

"5. La fixité de la température des corps à l'état sphéroïdal.

"Cela posé, voici la définition que je propose: un corps projeté sur une surface chaude est à l'état sphéroïdal quand il revêt la forme arrondie et qu'il se maintient sur cette surface au delà du rayon de sa sphère d'activité physique et chimique; alors il réfléchit le calorique rayonnant, et ses molécules sont, quant à la chaleur, dans un état d'équilibre stable, c'est-à-dire, à une température invariable, ou qui ne varie que dans des limites étroites." —Comptes Rendus, 6 Mars, 1848.

([79]) p. 89.—*Some facts relative to the Spheroidal state of Bodies, Fire Ordeal, Incombustible Man, &c.:* by P. H. Boutigny (d'Evreux), Philosophical Magazine, No. 233 (third series), p. 80. Comptes Rendus, May 14, 1849.

([80]) p. 90.—The theory of freezing mixtures is deduced from the doctrine of latent caloric. These are mixtures of saline substances which, at the common temperature, by their mutual chemical action, pass rapidly into the fluid form, or are capable of being rapidly dissolved in water, and, by this quick transition to fluidity, absorb caloric, and produce degrees of cold more or less intense.—Rev. Francis Lunn, *On Heat:* Encyclopædia Metropolitana.

([81]) p. 90. — *Propriéts de l'Acide Carboniqus liquide:* par M. Thilorier: Annales de Chimie, vol. lx. p. 427. *Solidification de l'Acide Carbonique:* Ibid. p. 432.

([82]) p. 91. — *On the Liquefaction and Solidification of Bodies generally existing as Gases:* by Michael Faraday, D. C. L., F. R. S., &c.; Philosophical Transactions, vol. cxxxvi. p. 155.

([83]) p. 92. — Burns, in one of his most natural and most pathetic letters.

([84]) p. 93. — "These — oxygen, hydrogen, nitrogen, and carbon — are the four bodies, in fact, which, becoming animated at the fire of the sun, the true torch of Prometheus, approve themselves upon the earth the eternal agents of organization, of sensation, of motion, and of thought." — Dumas, *Leçons de Philosophie Chimique*, p. 100, Paris, 1837.

([85]) p. 94. — It will be found in examining any of the works of the alchemists, — particularly those of Geber, *De inveniendi arte Auri et Argenti*, and his *De Alchemiâ;* Roger Bacon's *Opus Majus, or Alchymia Major;* Helvetius' *Brief of the Golden Calf;* or Basil Valentine's *Currus Triumphalis*, — that in the processes of transmutation, the solar light was supposed to be marvellously effective. In Boyle's *Skeptical Chemist*, the same idea will be found pervading it.

Amid all their errors, the alchemists were assiduous workmen, and to them we are indebted for numerous facts. Of them, and of their age, however, as contrasted with our own, Gibbon remarks: "Congenial to the avarice of the human heart, it was studied in China, as in Europe, with equal eagerness and equal success. The darkness of the Middle Ages insured a favourable reception to every tale of wonder; and the revival of learning gave new vigor to hope, and suggested more specious arts of deception. Philosophy, with the aid of experience, has at length banished the study of alchemy; and the present age, however desirous of riches, is content to seek them by the humbler means of commerce and industry." — *Decline and Fall*, vol. ii. p. 137.

([86]) p. 95. — On the two theories the following may be consulted: — Young, *Supplement to Encyclopædia Britannica*, article *Chromatics;* Fresnel, *Supplément à la Traduction Française de la 5ième édition du Traité de Chimie de Thomson*, par Riffault, Paris, 1822; Herschel's article, *Light*, in the Encyclopædia Metropolitana, and the French translation of it by Quetelet and Verhulst; Airy's *Tract on the Undulatory Theory*, in his Tracts, 2nd edition, Cambridge, 1831; Powell, *The Undulatory Theory applied to Dispersion*, &c., p. 184; Lloyd's

Lectures, Dublin, 1836–41; Cauchy, *Sur le Mouvement des Corps élastiques*, Mémoires de l'Institut, 1827, vol. ix. p. 114; *Théorie de la Lumière*, Ibid., vol. x. p. 293; M'Cullagh, *On Double Refraction*, Ibid., vol. xvi.; *Geometrical Propositions applied to the Wave Theory of Light*, Ibid., vol. xvii.; Sir David Brewster's papers in the Transactions of the Royal Society of Edinburgh, and the Philosophical Magazine.

(87) p. 96. — *Results of Astronomical Observations made during the years* 1834–38, *at the Cape of Good Hope, &c.* By Sir John Herschel, Bart., K.H., D.C.L., F.R.S. — "In the contemplation of the infinite, in number and in magnitude, the mind ever fails us. We stand appalled before this mighty spectre of boundless space, and faltering reason sinks under the load of its bursting conceptions. But, placed as we are on the great locomotive of our system, destined surely to complete at least one round of its ethereal course, and learning that we can make no apparent advance on our sidereal journey, we pant with new ardour for that distant bourne which we constantly approach without the possibility of reaching it. In feeling this disappointment and patiently bearing it, let us endeavour to realize the great truth from which it flows. It cannot occupy our mind without exalting and improving it." — North British Review.

(88) p. 98. — For examples of this, consult Graham's *Elements of Chemistry;* Brande's *Manual of Chemistry;* or, indeed, any work treating of the science. The formation of ink, by mixing two colourless solutions, one of gallic acid and another of sulphate of iron, may be taken as a familiar instance.

(89) p. 98. — Sir John Herschel, in his paper *On the Chemical Action of the Rays of the Solar Spectrum on preparations of Silver*, remarks, that "it may seem too hazardous to look for the cause of this very singular phenomenon in a real difference between the chemical agencies of those rays which issue from the central portion of the sun's disc, and those which, emanating from its borders, have undergone the absorptive action of a much greater depth of its atmosphere; and yet I confess myself somewhat at a loss what other cause to assign for it. It must suffice, however, to have thrown out the hint; remarking only, that I have other, and, I am disposed to think, decisive evidence (which will find its place elsewhere) of the existence of an absorptive solar atmosphere, extending beyond the luminous one. The breadth of the border, I should observe, is small, not exceeding 0.5 or $\frac{1}{7}$ part of the sun's radius, and this, from the circumstances of the experiment, must necessarily err in excess." — Philosophical Transactions, 1840.

(90) p. 99. — *Experiments and Observations on some Cases of Lines in the Prismatic Spectrum, produced by the passage of Light through Coloured Vapours and Gases, and from certain Coloured Flames:* by W. A. Millar, M.D., F.R.S., Professor of Chemistry in King's College, London. — Philosophical Magazine, vol. xxvii.

(91) p. 99. — *Report on the Mollusca and Radiata of the Ægean Sea, and on their distribution, considered as bearing on Geology.* By Edward Forbes, F.R.S., &c. — Reports of the British Association, vol. xii. Professor Forbes remarks: "A comparison of the testacea, and other animals of the lowest zones, with those of the higher, exhibits a very great distinction in the hues of the species, those of the depths being, for the most part, white or colourless, while those of the higher regions, in a great number of instances, exhibit brilliant combinations of colour. The results of an inquiry into this subject are as follows: —

"The majority of shells of the lowest zone are white or transparent; if tinted rose is the hue, a very few exhibit markings of another colour. In the seventh region, white species are also very abundant, though by no means forming a proportion so great as the eighth. Brownish red, the prevalent hue of the brachiopoda, also gives a character of colour to the fauna of this zone; the crustacea found in it are red. In the sixth zone the colours become brighter, reds and yellows prevailing, — generally, however, uniformly colouring the shell. In the fifth region many species are banded or clouded with various combinations of colours, and the number of white species has greatly diminished. In the fourth, purple hues are frequent, and contrasts of colour common. In the second and third, green and blue tints are met with, sometimes very vivid; but the gayest combinations of colour are seen in the littoral zone, as well as the most brilliant whites.

"The animals of Testacea, and the Radiata of the higher zones, are much more brilliantly coloured than those of the lower, where they are usually white, whatever the hue of the shell may be. Thus the genus *Trochus* is an example of a group of forms mostly presenting the most brilliant hues both of shell and animal; but whilst the animals of such species as inhabit the littoral zone are gaily chequered with many vivid hues, those of the greater depth, though their shells are almost as brightly covered as the coverings of their allies nearer the surface, have their animals, for the most part, of a uniform yellow or reddish hue, or else entirely white. The chief cause of this increase of intensity of colour as we ascend, is, doubtless, the increased amount of light above a certain depth." — p. 172.

([92]) p. 100. — *'Αμόρφωτα. On the Epipolic Dispersion of Light*, being a paper entitled, *On a case of Superficial Colour presented by a homogeneous liquid internally colourless.* By Sir J. F. W. Herschel, Bart., K. H., F.R.S., &c. — *An epipolarized beam of light*, (meaning thereby a beam which has once been transmitted through a quiniferous solution, and undergone its dispersing action,) *is incapable of further undergoing epipolic dispersion.* In proof of this the following experiment may be adduced.

Exp. A glass jar being filled with a quiniferous solution, a piece of plate glass was immersed in it vertically, so as to be entirely covered, and to present one face directly to the incident light. In this situation, when viewed by an eye almost perpendicularly over it, so as to graze either surface very obliquely, neither the anterior nor posterior face showed the slightest trace of epipolic colour. Now, the light, at its egress from the immersed glass, entered the liquid under precisely the same circumstances as that which, when traversing the anterior surface of the glass jar, underwent epipolic dispersion on first entering the liquid. It had, therefore, lost a property which it originally possessed, and could not, therefore, be considered *qualitatively* the same light. — Philosophical Transactions, vol. cxxxvi.

([93]) p. 101. — In connection with this view, the Newtonian theory should be consulted, for which see — *A Letter of* Mr. Isaac Newton, *Professor of the Mathematicks in the University of Cambridge; containing his new Theory about Light and Colours: sent by the Author to the Publisher from Cambridge, Feb.* 6, 1671 – 72, *in order to be communicated to the R. Society.*

([94]) p. 102. — In that admirable work, *The Physical Atlas* of Dr. Berghaus, of which a very complete edition by Alexander Keith Johnstone is published in this country, the following order of the distribution of plants is given: —

1. The region of palms and bananas	Equatorial zone.
2. Tree-ferns and figs . . .	Tropical zone.
3. Myrtles and laurels . . .	Sub-tropical zone.
4. Evergreen trees	Warm temperate zone.
5. European trees	Cold temperate zone.
6. Pines	Sub-arctic zone.
7. Rhododendrons	Arctic zone.
8. Alpine plants	Polar zone.

Consult Humboldt, *Essai sur la Géographie des Plantes*, Paris, 1807; *De Distributione Geographicâ Plantarum*, Paris, 1817. Schouw, *Grundzüge der Pflanzengeographie*. Lamouroux, *Géographie Physique*. *The Plant, a Biography:* by Schleiden; translated by Henfrey. *Physical Geography:* by Mrs. Somerville.

([95]) p. 103.—Frauenhofer's measurement of illuminating power is as follows:—

At the 22nd degree of the red	0.032
" 34th degree of the red	0.094
" 22nd degree of the orange	0.640
" 10th degree of the yellow	1.000
" 42nd degree of the yellow	0.480
" 2nd degree of the blue	0.170
" 16th degree of the indigo	0.031
" 43rd degree of the violet	0.0056

([96]) p. 103.—Herschel, *On the Action of Crystallized Bodies on Homogeneous Light, and on the causes of the deviation from Newton's scale in the tints which many of them develope on exposure to a polarized ray.*—Phil. Trans., vol. cx. p. 88.

([97]) p. 105.—*On the Nature of Light and Colours:* Lecture 39, in Young's *Lectures on Natural Philosophy*, Kelland's Edition, p. 373, and the authorities there quoted.

([98]) p. 106.—Brewster's *Optics:* Lardner's Cabinet Cyclopædia. Herschel, *On Light;* Encyclopædia Metropolitana.

([99]) p. 107.—Malus, *Sur une Propriété de la Lumière Réflêchie:* Mémoires d'Arcueil. Numerous memoirs by Sir David Brewster, in the Philosophical Transactions.

([100]) p. 108.—Bartholin, *On Iceland Crystals:* Copenhagen, 1669. *An Accompt of sundry Experiments made and communicated by that Learn'd Mathematician* Dr. Erasmus Bartholin, *upon a Chrystal-like Body sent to him out of Island:* in connection with which Dr. Matthias Paissenius writes: The observations of the excellent Bartholin upon the Island Chrystal are, indeed, considerable, as well as painful. We have here, also, made some tryals of it upon a piece he presented me with, which confirm his observations. Mean time we found it somewhat scissile and reducible by a knife into thin laminas or plates, which, when single, shew'd the object single, but laid upon one another shew'd it double; the two images appearing the more distant from one another, the greater the number was of those thin plates laid on one another. With submission to better judgements I think it to be a kind of Selenites. Some of our curious men here were of opinion that the Rhomboid figure proper to this stone was the cause of the appearances doubled thereby. But having tryed whether in other transparent bodies of the like figure the like would happen, we found no such thing in them, which made us suspect some peculiarity in the very Body of this Stone. — Phil. Trans. for 1670, vol. v.

([101]) p. 108. — *On the Application of the Laws of Circular Polarization to the Researches of Chemistry:* by M. Biot. — Nouvelles Annales du Muséum d'Histoire Naturelle, vol. iii., and Scientific Memoirs, vol. i. p. 600. *On Circular Polarization:* by Dr. Leeson. — Memoirs of the Chemical Society.

([102]) p. 109. — In Sir David Brewster's Treatise *On Optics*, chap. xviii., *On Polarization*, the best arrangements for a polarizing apparatus will be found described.

([103]) p. 109. — A beautiful application of this fact has recently been made by Professor Wheatstone, in the formation of an instrument for determining solar time by the polarization of the northern sky. *On a means of determining the apparent Solar Time by the diurnal changes of the Plane of Polarization at the North Pole of the Sky:* Report of the Eighteenth Meeting of the British Association.

([104]) p. 109. — *On the Polarization of the Chemical Rays of Light:* by John Sutherland, M.D., in which the author refers to the following experiment of M. J. E. Bérard: "I received the chemical rays directed into the plane of the meridian on an unsilvered glass, under an incidence of 35° 6′. The rays reflected by the first glass were received upon a second, under the same incidence. I found that when this was turned towards the south, the muriate of silver exposed to the invisible rays which it reflected was darkened in less than half an hour; whereas, when it was turned towards the west, the muriate of silver exposed in the place where the rays ought to have been reflected, was not darkened, although it was left exposed for two hours. It is consequently to be presumed that the chemical rays can undergo double refraction in traversing certain diaphanous bodies; and lastly, we may say that they enjoy the same physical properties as light in general." — Philosophical Magazine, vol xx.

Dr. Leeson has stated that Daguerreotype pictures can be taken more readily under the influence of polarized light, than by ordinary radiation.

([105]) p. 110. — *On the Magnetization of Light, and the Illumination of Magnetic Lines of Force:* by Michael Faraday, D.C.L., F.R.S. —

Philosophical Transactions, vol. cxxxvii. — The following remarks are to the point of doubt referred to in the text: "The magnetic forces do not act on the ray of light directly and without the intervention of matter, but through the mediation of the substance in which they and the ray have a simultaneous existence; the substances and the forces giving to and receiving from each other the power of acting on the light. This is shown by the non-action of a vacuum, of air or gases, and it is also further shown by the special degree in which different matters possess the property. That magnetic force acts upon the ray of light always with the same character of manner, and in the same direction, independent of the different varieties of substance, or their states of solid or liquid or their specific rotative force, shows that the magnetic force and the light have a direct relation; but that substances are necessary, and that these act in different degrees, shows that the magnetism and the light act on each other through the intervention of the matter. Recognizing or perceiving *matter* only by its powers, and knowing nothing of any imaginary nucleus abstract from the idea of these powers, the phenomena described must strengthen my inclination to trust in the views I have advanced in reference to its nature." — Phil. Mag. vol. xxiv.

(106) p. 111. — The invention of the camera obscura certainly belongs to Giambattista Porta, and is described in his *Magia Naturalis, sive de Miraculis Rerum Naturalium, Libri Viginti*; Antwerp, 1561. An English translation made in 1658 exists, but I have not seen it.

Hooke, in one of the earliest volumes of the Philosophical Transactions, describes, as new, many of the phenomena mentioned by Porta, and particularly the images of the dark chamber.

(107) p. 112. — Herschel, *On Light*. — Encyclopædia Metropolitana.

(108) p. 114. — "I would here observe that a consideration of many such phenomena (the obliteration and revival of photographic drawings) has led me to regard it as not impossible that the retina itself may be *photographically* impressed by strong light, and that some at least of the phenomena of visual spectra and secondary colours may arise from the sensorial perception of actual changes in progress in the physical state of that organ itself subsequent to the cessation of the direct stimulant." — *On the Action of the Rays of the Solar Spectrum on Vegetable Colours, &c.*: by Sir J. F. W. Herschel, Bart.

(109) p. 116. — Dumeril.

(110) p. 119. — *Theory of Colours*: by Goethe; translated by Eastlake.

([111]) p. 120.—See Tuckey's Narrative of the Expedition of the Zaire.

([112]) p. 120.—The most complete examination of this subject will be found in two Memoirs:—

1. *Experiments and observations on the light which is spontaneously emitted with some degree of permanency from various bodies.*—Phil. Trans. vol. xc.

2. *A continuation of the above, with some experiments and observations on solar light, when imbibed by Canton's phosphorus:* by Nathaniel Hulm, M.D.—Phil. Trans., vol. xci.; and in the *Monograph of the British Naked-eyed Medusæ*, by Professor Edward Forbes, (published for the Ray Society.) See Wilson's note to the account of *Pennalata phosphorea* in Johnston's Zoophytes, 2nd edition.

([113]) p. 120.—*A general Outline of the Animal Kingdom:* by Thomas Rymer Jones, F.L.S.—Acalephæ, p. 64. *Lettre à M. Dumas sur la Phosphorescence des Vers luisants:* par M. Ch. Matteucci.—Annales de Chimie, vol. ix. p. 71, 1843.

([114]) p. 120.—*Phosphorescence of the Diamond:* by M. Reiss (Revue Scientifique et Industrielle, vol. xxiii. p. 185).—"The diamond, phosphorescent by insulation, lost rapidly its phosphorescence when submitted to the action of the red rays of the solar spectrum. On the contrary, the blue rays are those which render the diamond the most luminous in the dark. It is probable that the phosphorescence produced by heat is equally diminished by the action of the red rays of the solar spectrum." Giovanni Battista Beccaria published his experiments in 1769.—See Priestley's *History of Electricity;* and *On the Effects of Electricity upon Minerals which are Phosphorescent by Heat;* and *Further Experiments on the communication of Phosphorescence and Colour to bodies of Electricity:* by Thomas J. Pearsall.—Journal of the Royal Institution of Great Britain, Oct. 1830, Feb. 1831.—These two memoirs contain the most complete set of experiments on this subject which have yet been made; see *Placidus Heinrich, Phosphorescenz der Körper*, vol. iv.; Gmelin's *Handbach der Chemie* part 1;—*On the Phosphorescence of Minerals*, Brewster. Edinburgh Philosophical Journal, vol. i. p. 137;—*The Aërial Noctiluca, or some New Phenomena, and a process of a factitious self-shining substance.* Boyle's Works, vol. iv.

([115]) p. 121.—*Des Effets produits sur les corps par les Rayons Solaires:* par M. Edmond Becquerel.—Annales de Chimie, vol. ix. p. 257. 1843.

M. Becquerel has applied the term *phosphorogénique* to those rays producing phosphorescence.

([116]) p. 127. — See *Researches on Light*, by the Author. — Reference to any of the works of the alchemists will prove the prevalence of the idea expressed in the text. We find that gold was considered to be always under the influence of light and solar heat. — "It is said of gold that it waxeth cold towards daylight, insomuch that they who wear rings of it may perceive when the day is ready to dawn." — *Speculum Mundi, or a Glass representing the face of the World.* Cambridge, 1643.

([117]) p. 128. — Daguerre's Report to the Academy of Sciences: *Le Daguerréotype Historique, et description des procédés du Daguerréotype et du Diorama* (Paris, 1839); particularly the description of *Heliography*, by M. Niepce; see also the letters by Niepce, published for the first time in *Researches on Light*.

([118]) p. 131. — "If a solution of peroxalate of iron be kept in a dark place, or if it be exposed to 212° of Fahr. for several hours, it does not undergo any sensible change in its physical properties, nor does it exhibit any phenomenon which may be considered as the result of any elementary action.

"If, however, it be exposed to the influence of solar light on a glass vessel provided with a tube, the concentrated solution of oxalate of iron soon presents a very interesting phenomenon: in a short time the solution receiving the solar rays, developes an infinite number of bubbles of gas, which rise in the liquor with increasing rapidity, and give the solution the appearance of a syrup undergoing strong fermentation. This ebullition always becomes stronger, and almost tumultuous, when an unpolished glass tube is immersed in it with a small piece of wood; the liquid itself is afterwards thrown into ascending and descending currents, becomes gradually yellowish, turbid, and eventually precipitates protoxalate of iron, in the form of small brilliant crystals of a lemon-yellow colour, gas continuing to evolve." *Chemical action of light, and formation of Humboldtine by it:* Phil. Mag. 1832, second series. — "When a solution of platinum in nitro-muriatic acid, in which the excess of acid has been neutralized by the addition of lime, and which has been well cleared by filtration, is mixed with lime-water, in the dark no precipitation to any considerable extent takes place for a long while, — indeed, none whatever, though after very long standing a slight flocky sediment is formed, after which, the action is arrested entirely. But if the mixture, either freshly made or when cleared by subsidence of this sediment, is exposed to sunshine, it instantly becomes milky, and a copious formation of a white precipitate (or a pale yellow one, if the platinic solution be in excess) takes place, which subsides quickly and is easily collected. The same takes place more slowly in cloudy daylight." — *On the action of light in determining the precipitation of Muriate of Platinum by Lime-water;* being an extract from Sir John G. W. Herschel, K.H., F.R.S., &c., to Dr. Daubeny. — Phil. Mag. 1832.

([119]) p. 132. — *On a change produced by Exposure to the Beams of the Sun, in the properties of an elementary substance,* by Professor Draper: *On the changes which bodies undergo in the dark,* by Robert Hunt: Report of the thirteenth meeting of the British Association, vol. xii. — *Description of the Tithonometer, an instrument for measuring the chemical force of the Indigo-tithonic rays:* by J. W. Draper, M. D. — Philosophical Magazine, Dec. 1843, vol. xxiii.

([120]) p. 133. — For several illustrations of this remarkable phenomenon, see *On the Action of the Rays of the Solar Spectrum on Vegetable Colours, and on some new Photographic Processes:* by Sir John F. W. Herschel, Bart., K.H., F.R.S. — Phil. Trans. June, 1842, vol. cxxxiii.; *On certain improvements on Photographic Processes described in a former communication, and on the Parathermic Rays of the Solar Spectrum:* by Sir John F. W. Herschel, Bart., K.H., F.R.S., &c., in a letter addressed to S. Hunter Christie. — Phil. Trans. 1843, vol. cxxxiv.

([121]) p. 133. — Sir J. F. W. Herschel; see also *Researches on Light,* by the Author.

([122]) p. 134. — Attention has been directed to the protecting action of certain rays of the spectrum by Sir John Herschel and others. See the Eighteenth Report of the British Association for an experiment by the Author, in which it was proved that all the LIGHT rays protected photographic papers from chemical change, and, therefore, convincingly show that light and actinism were not similar powers.

([123]) p. 135. — "Having noticed, one densely foggy day, that the disc of the sun was of a deep red colour, I directed my apparatus towards it. After ten seconds of exposure, I put the prepared plate in the mercury box, and I obtained a round image perfectly black; — the sun had produced no photogenic effect. In another experiment, I left the plate operating for twenty minutes; — the sun had passed over a certain space of the plate, and there resulted an image seven or eight times the sun's diameter in length; it was black throughout, so that it was evident, wherever the red disc of the sun had passed, not only was there a want of photogenic action, but the red rays had destroyed the effect produced previous to the sun's passage,

I repeated these experiments during several days successively, operating with a sun of different tints of red and yellow. These different tints produced nearly the same effect; wherever the sun had passed, there existed a black band." — Mr. Claudet, *On different properties of Solar Radiation, modified by coloured glass media, &c.*; Phil. Trans. 1847. Part 2.

([124]) p. 136. — "It may also be observed that the rays effective in destroying a given tint are, in a great many cases, those whose union produces a colour complementary to the tint destroyed, or at least one belonging to that class of colours to which such complementary tint may be referred. For example, yellows tending towards orange are destroyed with more energy by the blue rays; blue by the red, orange, and yellow rays; purples and pinks by yellow and green rays." — Sir J. F. W. Herschel, *On the action of the rays of the Solar Spectrum on Vegetable Colours*: Phil. Trans., vol. cxxxiii. 1842.

([125]) p. 136. — The following memoirs and works are necessary to a complete history of the inquiry: *Experiments and observations relating to various branches of natural philosophy, with a continuation of the observations on air*: by Dr. Priestley, London, 1779. *Mémoires Physico-chimiques, &c.*: by J. Senebier. *Expériences sur les végétaux*, by De la Ville: Paris, 1782; and Phil. Trans. 1782. *Observations sur les expériences de M. Ingenhousz*: by De la Ville; Ros. obs. 23, 290. *Expériences propres à développer les effets de la lumière sur certaines plantes*: by Tessier; Mém. de l'Ac. des Sc. de Paris, 1783, p. 132; Licht. Mag. iv. 4, 146. *Sur la vertu de l'eau impregnée d'air fixe pour en obtenir, par le moyen des plantes et de la lumière du soleil, de l'air déphlogistiqué*: by Ingenhousz; Roz. obs. 24, 337. *Expériences sur l'action de la lumière solaire dans la végétation*: by Senebier; Genève et Paris, 1788, p. 61. *Extrait des expériences de M. Senebier sur l'action de la lumière solaire dans la végétation*: by Hasenfratz; Ann. Chim. iii. 2nd ser. 266. *Expériences relatives à l'influence de la lumière sur quelques végétaux*: by De Candolle; Jour. de Ph. lii. 124; Voigt's Mag. ii. 483; Gilb. Ann. xiii. 372; Mém. des Sav. Etr. i. 329. *Recherches chimiques sur la végétation*: by Saussure; Ann. Chim. l. 225; Jour. de Ph. lvii. p. 393; Gilb. Ann. xviii. 208. *Recherches sur la respiration des plantes exposées à la lumière du soleil*: by Ruhland; Ann. Ch. Ph. iii. 411; Jour. de Ph. 1816. *On the action of light upon plants, and of plants upon the atmosphere*: by Dr. Daubeny; Phil. Trans. cxxvii. January, 1836. *On the action of yellow light in producing the green colour, and of indigo light on the movements of plants*: by P. Gardner; Phil. Mag. xxiv.; Bibl. Univ. xlix. p. 376, and lii. p. 381. *On the influence of light on plants*: by R. Hunt; Phil. Mag. xxiv. p. 96; Bibl. Univ.

xlix. p. 383; Athen. 1844. *Note on the decomposition of carbonic acid by the leaves of plants, under the influence of yellow light:* by Draper; Phil. Mag. xxv. p. 169. *On the action of the yellow rays of light on vegetation:* by Harkness; Phil. Mag. xxv. p. 339. *Influence des rayons solaires sur la végétation:* by Zantedeschi; Inst. No. 541, p. 157.

(126) p. 136. — Sir John Herschel's Memoirs, already referred to; and *Reports on the influence of the Solar Rays on the growth of Plants*, by Robert Hunt: Report of the British Association for the Advancement of Science, for 1847.

(127) p. 137. — *Memoir on the Constitution of the Solar Spectrum*, presented at the meeting of the Academy of Sciences, 1842, by M. Edmond Becquerel; *Des effets produits sur les corps par les rayons solaires*, par M. Edmond Becquerel, aide au Muséum d'Histoire Naturelle: Mémoire présenté à l'Académie des Sciences, le 23 Octobre, 1843. — "Dans le courant de ce mémoire, j'ai employé les noms de rayons lumineux, chimiques, et phosphorogéniqués, pour désigner, dans chaque cas, la portion des rayons solaires qui agit pour prôduire, en particulier, les effets lumineux, chimiques, et phosphorogéniques; mais cela est sans préjudice de l'opinion que je viens d'émettre touchant l'existence d'un seul et méme rayonnement."

"My reply is this," says M. Arago, in his paper entitled *Considerations relative to the action of Light:* "it is by no means proved that the photogenic modifications of sensitive substances result from the action of solar light itself. The modifications are, perhaps, engendered by invisible radiations mixed with light properly so called, proceeding with it, and being similarly refracted. In this case, the experiment would prove not only that the spectrum formed by these invisible rays is not continuous, that there are solutions of continuity as in the visible spectrum, but also that in the two superposed spectra these solutions correspond exactly. This would be one of the most curious, one of the most strange results of physics." — Taylor's Scientific Memoirs.

(128) p. 138. — The chemical evidence of this will be found in Sir John Herschel's Memoir *On the Solar Spectrum*, and particularly as exemplified in the changes produced on the tartrate of silver. Similar influences are described as observed on a Daguerreotype plate, in a paper entitled *Experiments and Observations on Light which has permeated coloured media, and on the Chemical Action of the Solar Spectrum:* by Robert Hunt. — Philosophical Magazine, vol. xxvi. 1840.

(129) p. 139. — This peculiar continuance of an effect has frequently been observed in many of the photographic processes. In a note

to a memoir *On certain improvements in Photographic processes*, Sir John Herschel thus refers to this property: "The excitement is produced on such paper by the ordinary moisture of the atmosphere, and goes on slowly working its effect in the dark, apparently without other limit than is afforded by the supply of ingredients present. In the case of silver it ultimately produces a perfect *silvering* of all the sunned portions. Very singular and beautiful photographs, having much resemblance to Daguerreotype pictures, are thus produced; the negative character changing by keeping, and by quite insensible gradations to positive, and the shades exhibiting a most singular *chatoyant* change of colour from ruddy-brown to black, when held more or less obliquely. No doubt, also, gold pictures with the metallic lustre might be obtained by the same process, though I have not tried the experiment."

([130]) p. 140. — The details of this curious subject may be studied in the following memoir and communications: *On vision and the action of light on all bodies:* by Professor Ludwig Moser, of Königsberg; from Poggendorff's Annalen, vol. lvi. p. 177, No. 6, 1845. *Some remarks on Invisible Light:* by Professor Ludwig Moser, of Königsberg; from Poggendorff's Annalen, vol. lvi. p. 569, No. 8. *On the power which light possesses of becoming latent:* by Professor Ludwig Moser, of Königsberg; from Poggendorff's Annalen, vol. lvii., No. 9, p. 1, 1842. *On certain spectral appearances, and on the discovery of latent light:* by J. W. Draper, M.D., Professor of Chemistry in the University of New York; Phil. Mag. p. 348, Nov. 1842. *On a new imponderable substance, and on a class of chemical rays analogous to the rays of dark heat:* by Professor Draper; Phil. Mag., Dec. 1842. *On the action of the rays of the solar spectrum on the Daguerreotype plate:* by Sir. J. F. W. Herschel, Bart.; Phil. Mag., Feb. 1843. See remarks in this paper on the use which Moser has made of coloured glasses; also a communication, by Professor Draper, *On the rapid Detithonizing power of certain gases and vapours, and on an instantaneous means of producing spectral appearances:* Phil. Mag., March, 1843; and *On the causes which concur in the production of the images of Moser:* Comptes Rendus, Nov. 1842, See *Scientific Memoirs*, vol. iii.

([131]) p. 140. — This fact was first observed by myself, and described in the paper already referred to, Philosophical Magazine, vol. xxii. p. 270. It does not, however, appear to have attracted the attention of any other observer.

([132]) p. 141. — *On Thermography, or the Art of copying Engravings or any printed characters from paper or plates of metal, and on the recent discovery of Moser, relative to the formation of images in the dark,* by

Robert Hunt: Reports of the Royal Cornwall Polytechnic Society for 1842, and Philosophical Magazine, vol. xxi. p. 462. — *On the Spectral images of M. Moser*, by Robert Hunt: Philosophical Magazine, vol. xxiii. p. 415.

([133]) p. 141. — *Catalytic force, or attraction of surface concerned in the diffusive power of gases: an occult energy or power in saturated saline solutions:* Prater. — Mechanic's Magazine, vol. xlv. p. 106. *Ueber elektrische Abbildungen:* by G. Karsten. — Poggendorff's Annalen, vol. lvii. p. 402. — Melloni and Brewster may be consulted for much that is most remarkable connected with radiation from coloured surfaces.

([134]) p. 142. — Cornelius Agrippa is said to have possessed such a mirror. The Chinese make mirrors which, when placed in a particular light, show upon their polished faces the pattern on the back of the metal, although it is invisible in every other position. This is effected by giving different degrees of hardness to the various parts of the metal. In *Natural Magic*, by Sir David Brewster, several curious experiments belonging to this class are named.

([135]) p. 146.—*Traité de Physique:* M. Biot, vol. vii. Becquerel: Annales de Chimie, vol. xlvi.—xlix. Faraday's *Experimental Researches in Electricity*, 2 vols., 1830—1844. *A Speculation touching Electric Conduction and the Nature of Matter:* by Michael Faraday, D. C. L., F. R. S.; Philosophical Magazine, vol. xxiv., 1836. *Objections to the theories severally of Franklin, Dufay, and Ampère, with an attempt to explain Electrical Phenomena by statical or undulatory polarization:* by Robert Hare, M. D., Emeritus Professor of Chemistry in the University of Pennsylvania.

([136]) p. 148. — "A good piece of gutta percha will insulate as well as an equal piece of shell-lac, whether it be in the form of sheet, or rod, or filament; but being tough and flexible when cold, as well as soft when hot, it will serve better than shell-lac in many cases where the brittleness of the latter is an inconvenience. Thus it makes very good handles for carriers of electricity in experiments on induction; not being liable to fracture in the form of thin band or string, it makes an excellent insulating suspender; a piece of it in sheet makes a most convenient insulating basis for anything placed on it. It forms excellent insulating plugs for the stems of gold-leaf electrometers, when they pass through sheltering tubes, and larger plugs form good insulating feet for electrical arrangements; cylinders of it, half an inch or more in diameter, have great stiffness, and form excellent insulating pillars. In these and in other ways, its power as an insulator may be useful." — *On the use of Gutta Percha in*

Electrical Insulation: by Dr. Faraday; Philosophical Magazine, March, 1848.

The following deductions have been given by Faraday, in his *Researches in Electricity*, a work of most extraordinary merit, being one of the most perfect examples of fine inductive philosophy which we possess in the English language: —

"All bodies conduct electricity in the same manner from metals to lacs and gases, but in very different degrees.

"Conducting power is in some bodies powerfully increased by heat, and in others diminished, yet without one perceiving any accompanying essential electrical difference, either in the bodies, or in the change occasioned by the electricity conducted.

"A numerous class of bodies insulating electricity of low intensity, when solid, conduct it very freely when fluid, and are then decomposed by it.

"But there are many fluid bodies which do not sensibly conduct electricity of this low intensity; there are some which conduct it and are not decomposed; nor is fluidity essential to decomposition.

"There are but two bodies (sulphuret of silver and fluoride of lead) which, insulating a voltaic current when solid, and conducting it when fluid, are not decomposed in the latter case.

"There is no strict electrical distinction of conduction which can as yet be drawn between bodies supposed to be elementary, and those known to be compounds."

(137) p. 149. — Faraday's *Speculation on the Nature of Matter*, already referred to.

(138) p. 149. — *Experimental Researches:* by Dr. Faraday. *Chemical Decomposition*, p. 151.

(139) p. 149. — Karsten; Poggendorff's *Annalen*, vol. lvii.

(140) p. 152. — *Traité Expérimental de l'Electricité et du Magnétisme:* Becquerel, 1834. Priestley's *Introduction to Electricity*. *On Electricity in Equilibrium:* Dr. Young's Lectures.

(141) p. 152. — Faraday's *Experimental Researches on Electricity*. This philosopher has shown, by the most conclusive experiments, "that the electricity which decomposes, and that which is evolved by the decomposition of, a certain quantity of matter, are alike. What an enormous quantity of electricity, therefore, is required for the decomposition of a single grain of water! We have already seen that it must be in quantity sufficient to sustain a platinum wire $\frac{1}{104}$ of an inch in thickness, red hot, in contact with the air, for three minutes and three quarters. It would appear that 800,000 charges of a Leyden battery, charged by thirty turns of a very large and powerful plate machine, in full action — a quantity sufficient, if

passed at once through the head of a rat or cat, to have killed it as by a flash of lightning — are necessary to supply electricity sufficient to decompose a single grain of water; or, if I am right, to equal the quantity of electricity which is naturally associated with the elements of that grain of water, endowing them with their mutual chemical affinity."

([142]) p. 152. — *Experimental Researches:* Faraday.

([143]) p. 155. — The appearance of acid and alkaline matter, in water, acted on by a current of electricity, at the opposite electrified metallic surfaces, was observed in the first chemical experiments made with the column of Volta — (see Nicholson's Journal, vol. iv. p. 183, and vol. iv. p. 261, for Mr. Cruickshank's Experiments; and Annales de Chimie, tom. xxxvii. p. 233, for those of M. Desormes:) *On some Chemical Agencies in Electricity:* by Sir Humphrey Davy. — Philosophical Transactions for 1807. The various theories of electro-chemical decomposition are carefully stated by Faraday, in his fifth series of *Experimental Researches on Electricity*, in which he thus states his own views: "It appears to me that the effect is produced by an *internal corpuscular action* exerted according to the direction of the electric current, and that it is due to a force either *superadded to* or *giving direction to the ordinary chemical affinity* of the bodies present. The body under decomposition may be considered as a mass of acting particles, all those which are included in the course of the electric current contributing to the final effect; and it is because the ordinary chemical affinity is relieved, weakened, or partly neutralized by the influence of the electric current in one direction parallel to the course of the latter, and strengthened or added to in the opposite direction, that the combining particles have a tendency to pass in opposite courses."

([144]) p. 157. — "This capital discovery (chemical decomposition by electricity) appears to have been made in the first instance by Messrs. Nicholson and Carlisle, who observed the decomposition of water so produced. It was speedily followed up by the still more important one of Berzelius and Hisinger, who ascertained it as a general law, that, in all the decompositions so effected, the acids and oxygen become transferred and accumulated around the positive, and hydrogen, metals, and alkalies around the negative, pole of a voltaic circuit; being transferred in an invisible, and, as it were, a latent or torpid state, by the action of the electric current, through considerable spaces, and even through large quantities of water or other liquids, again to reappear with all their properties at their appropriate resting-places."—*Discourse on the Study of Natural Philosophy*: by Sir John Herschel, Bart., F.R.S.

([145]) p. 157. — Numerous beautiful illustrations of this fact will be found in Becquerel's *Traité Expérimental de l'Electricité et du Magnétisme.*

([146]) p. 158. — See *Le Feu élémentaire* of L'Abbé Nollet; Leçons de Physique, tom. vi. p. 252; *Du Pouvoir thermo-électrique*, by M. Becquerel — Annales de Chimie, vol. xli. p. 353; also, a Memoir by Nobili, Bibliothèque Universelle, vol. xxxvii. p. 15; *Experimental Contributions towards the theory of Thermo-Electricity*, by Mr. J. Prideaux — Philosophical Magazine, vol. iii., Third Series; *On the Thermo-Magnetism of Homogeneous Bodies, with illustrative experiments*, by Mr. William Sturgeon — Philosophical Magazine, vol. x. p. 1–116, New Series. Botto made magnets and obtained chemical decomposition. Antinori produced the spark. Mr. Watkins heated a wire in Harris's Thermo-Electrometer.

([147]) p. 158. — A very ingenious application of the knowledge of this fact was suggested by Mr. Solly, by which the heat of a furnace could be constantly registered at a very considerable distance from it. See *Description of an Electric Thermometer;* by E. Solly, Junr., Esq. — Philosophical Magazine, vol. xx. p. 391. New Series.

([148]) p. 159. — Humboldt; *Personal Narrative*, chap. xvii. — Annales de Chimie, vol. xiv. p. 15.

([149]) p. 159. — *Experimental Researches on Electricity.* Series xv. Consult Sir Humphrey Davy: *An Account of some Experiments on the Torpedo.* — Philosophical Transactions, 1829, p. 15. John Davy, M.D., F.R.S.: *An Account of some Experiments and Observations on the Torpedo*, ibid., 1832, p. 259; and the same author's *Observations on the Torpedo, with an account of some Additional Experiments on its Electricity;* and Matteucci, Bibliothèque Universelle, 1837, vol. xii. p. 174.

([150]) p. 161. — *On Lightning Conductors*, by Sir William Snow Harris; *Observations on the Action of Lightning Conductors*, by W. Snow Harris, Esq., F.R.S. — London Electrical Society's Transactions. Numerous valuable papers *On Electricity*, by Sir William Harris, will be found in the Philosophical Transactions.

([151]) p. 162. — Adopting, to a certain extent, this view, Faraday, in his *Electrical Nomenclature*, proposed for the word pole to substitute *anode* (ἄνω, *upwards*, and ὁδὸς, *a way*), the way which the sun rises; and *cathode* (κατὰ, *downwards*, and ὁδὸς, *a way*), the way which the sun sets. The hypothesis belongs essentially to Ampère. *Objections to the Theories severally of Franklin, Dufay, and Ampère, with an effort to explain Electrical Phenomena by statical or undulatory Polarization*, by Robert Hare, M.D., Pennsylvania, will well repay an attentive perusal.

(152) p. 162. — *Inquiry into the laws of the Vital Functions.* — Philosophical Transactions, 1815, 1822; *Some Observations relating to the Functions of Digestion*, ibid., 1829; *On the Powers of which the Functions of Life in the more perfect animals depend, and on the manner in which they are associated in the production of their more complicated results*, by A. P. W. Philip, M.D., F.R.S., L. & E. — The following extract from the last-quoted of Dr. Philip's Memoirs, will give a general view of the conclusions of that eminent physiologist: "With respect to the nature of the powers of the living animal which we have been considering, the sensorial and muscular powers, and the powers peculiar to living blood, we have found belong to the living animal alone, all their peculiar properties being the properties of life. The functions of life may be divided into two classes, those which are effected by the properties of this principle alone, and those, by far the most numerous class, which result from the co-operation of these properties with those of the principles which operate in inanimate nature. The nervous power we have found to be a modification of one of the latter principles, because it can exist in other textures than those to which it belongs in the living animal, and we can substitute for it one of those principles without disturbing the functions of life.

"Late discoveries have been gradually evincing how far more extensive than was supposed, even a few years ago, is the dominion of electricity. Magnetism, chemical affinity, and (I believe, from the facts stated in the foregoing paper, it will be impossible to avoid the conclusion) the nervous influence, the leading power in the vital functions of the animal frame, properly so called, appear all of them to be modifications of this apparently universal agent; for I may add we have already some glimpses of its still more extensive dominion."

Refer to Dr. Read's papers.

(153) p. 162. — *Electro-physiological Researches:* by Signor Carlo Matteucci; Phil. Trans. 1845, p. 293, and subsequent years.

(154) p. 163. — Electro-Biology: by Alfred Smee, Esq.

(155) p. 165. — *Observations of Electric Currents in Vegetable Structures:* by Golding Bird, Esq., F.L.S.; Magazine of Natural History, vol. x. p. 240. In this paper, Dr. Bird remarks that his experiments lead to the conclusion that vegetables cannot become so charged with electricity as to afford a spark; that electrical currents of feeble tension are always circulating in vegetable tissues; and that electrical currents are developed during germination from chemical action.

([156]) p. 166. — *On Mineral Veins:* by Robert Were Fox, Esq.; Fourth Report of the Royal Cornwall Polytechnic Society. *On the Electro-magnetic Properties of Metalliferous Veins in the mines of Cornwall:* by Robert W. Fox, Esq.; Phil. Trans., 1830, p. 399.

([157]) p. 167. — *Experiments and Observations on the Electricity of Mineral Veins:* by Robert Hunt and John Phillips; Reports of the Royal Cornwall Polytechnic Society for 1841–42. *On the Electricity of Mineral Veins:* by Mr. John Arthur Phillips; Ibid., 1843.

([158]) p. 167. — In the lead lodes of *Lagylas* and *Frongoch*, electrical currents were detected by Mr. Fox, but none in those of *South Mold* and *Milwr*, in Flintshire: Cornwall Geological Transactions, vol. iv. In the lead veins of *Coldberry* and *Skeers* in Teasdale, Durham, the currents detected were very feeble: Reports of the Bristol Association, 1838. Von Strombeck could detect no electric currents in the veins worked in the clay slate near Saint Goar, on the Rhine: Archiv für Mineralogie, Geognosie, &c., von Dr. C. J. B. Karsten, 1833. Professor Reich, however, obtained very decided results at *Frisch Glück*, *Neue Hoffnung Gottlob*, and in other mineral veins in the mining districts of Saxony: Edinburgh New Philosophical Journal, vol. xxviii. 1839. The irregularities are to be explained by the presence or absence of chemical excitation.

([159]) p. 167. — This was remarkably the case at *Huel Sparnon*, near Redruth, where the cobalt was discovered between two portions of a dislocated lode; and the same was observed by Mr. Percival Johnson, in a small mine worked for nickel, near St Austell.

([160]) p. 167. — *On the process used for obtaining artificial veins in clay:* by T. B. Jordan; Sixth Annual Report of the Royal Cornwall Polytechnic Society. See also my memoir already referred to, in the Memoirs of the Geological Survey and Museum of Practical Geology, vol. i.

([161]) p. 168. — See Becquerel, *Traité Expérimental de l'Electricité, &c.* *Electrical Experiments on the formation of Artificial Crystals:* by Andrew Crosse, Esq.; British Association Reports, vol. v., 1836. The lamination of clay and other substances is described in my memoir referred to, Note 155.

([162]) p. 168. — Report on the Geology of Cornwall, Devon, and West Somerset, by Sir Henry T. De la Beche: *Theoretical observations on the formation and filling of Mineral Veins and Common Faults,* p. 349.

([163]) p. 168. — The following analysis of waters from deep mines were made by me in 1840, and, with many others, published in the Reports of the Royal Cornwall Polytechnic Society.

Consolidated mines, Gwennap, Cornwall.

	In 1000 grains of water.
Muriate of soda	1.5
Sulphate of lime	.5
Sulphate of iron	.15
Sulphate of copper	1.25
Alumina	.3
Total	3.7

United Mines, Gwennap.

	In 1000 grains of water.
Muriate of soda	1.10
Muriate of lime	.15
Sulphate of soda	.50
Sulphate of lime	1.5
Sulphate of iron	.75
Alumina	.5
Silica	.15
Total	4.65

Great St. George.

Muriate of soda	1.35
Sulphate of lime	.74
Carbonate of iron	.70
Alumina	.50
Carbonate of lime	.10
Total	3.4

([164]) p. 170. — The discovery of the electrotype has been disputed, as all valuable discoveries are. Without, however, at all disparaging the merits of what had been done by Mr. Jordan, I am satisfied, after the most careful search, that the first person who really employed electro-chemical action for the precipitation of metals in an ornamental form, was Mr Spencer of Liverpool.

([165]) p. 170. — See Spencer, *Instructions for the Multiplication of works of Art in Metal by Voltaic Electricity. Novelties in Experimental Science:* Griffin, Glasgow. *Elements of Electro-metallurgy;* by Alfred Smee, Esq.

([166]) p. 170. — The magneto-electrical machine is employed in Birmingham for this purpose; but I am informed by Messrs. Elkington that they do not find it economical, or rather that the electro-precipitation is carried on too slowly.

([167]) p. 170.—This has been done by Mr. Robert Were Fox, at a mine near Falmouth. By connecting two copper wires with two lodes, and bringing them, at the surface, into a cell containing a solution of sulphate of copper, this gentleman obtained an electrotype copy of an engraved copper-plate.

([168]) p. 173.—This is stated to have been most effectually accomplished by Mr. Bain.

([169]) p. 175.—*Treatise on Magnetism*, by Sir David Brewster. *Cosmos: a Sketch of a Physical description of the Universe*, by Alexander Von Humboldt.—Sabine's Translation, vol. ii. p. 268.

([170]) p. 179.—*Expérience Electro-Magnétique:* par M. Œrsted.—Annales de Chimie, vol. xxii. p. 201. De la Rive, *Recherches sur la Distribution de l'Electricité dyn. dans les Corps.*—Genève, 1825.

([171]) p. 179.—*On the Magnetic power of Soft Iron:* by Mr. Watkins.—Philosophical Transactions, 1833.

([172]) p. 180.—Cavallo, *On Magnetism.*—Cavallo was the first who noticed the influence of heat on magnetism. Consult *On the anomalous Magnetic Action of hot Iron between the white and blood-red heat:* by Peter Barlow, Esq.—Philosophical Transactions, 1822, p. 124. *Treatise on Magnetism:* by Barlow.—Encyclopædia Metropolitana.

([173]) p. 181.—"The foundation of our researches is the assumption that the terrestrial magnetic force is the collective action of all the magnetized particles of the earth's mass. We represent to ourselves magnetization as the separation of the magnetic fluids. Admitting the representation, the mode of action of the fluids (repulsion of similar, and attraction of dissimilar, particles inversely as the square of the distance) belongs to the number of established truths. No alteration in the results would be caused by changing this mode of representation for that of Ampère, whereby, instead of magnetic fluids, magnetism is held to consist in constant galvanic currents in the minutest particles of bodies. Nor would it occasion a difference if the terrestrial magnetism were ascribed to a mixed origin, as proceeding partly from the separation of the magnetic fluids in the earth, and partly from galvanic currents in the same; inasmuch as it is known that for each galvanic current may be substituted such a given distribution of the magnetic fluids in a surface bounded by the current, as would exercise in each point of external space precisely the same magnetic action as would be produced by the galvanic current itself."—*General Theory of Terrestrial Magnetism:* by Professor Carl Friedrich Gauss, of the University of Göttingen.—Scientific Memoirs, vol. ii. p. 188.

([174]) p. 183. — Hansteen: *Untersuchungen über den Magnetismus der Erde.* Christiania, 1819. Humboldt: *Exposé des Variations Magnétiques.* — Gilbert's Annales. Brewster's Magnetism: Encyclopædia Metropolitana.

([175]) p. 184. — Hansteen; as above.

([176]) p. 184. — *On the effects of temperature on the intensity of magnetic forces, and on the diurnal variations of the terrestrial magnetic intensity:* by Samuel Hunter Christie, Esq. — Philosophical Transactions, vol. cxv. 1825.

([177]) p. 185. — It has been observed by Mr. Barlow, in England, and some eminent observers in Austria, that an electric current constantly traverses the wires of the electric telegraph wherever there are two earth connections.

([178]) p. 185. — *Meteorological Observations and Essays:* by Dr. Dalton. *On the Height of the Aurora Borealis above the surface of the Earth:* by John Dalton, F.R.S. — Philosophical Transactions, vol. cxiv. p. 291.

([179]) p. 185. — Arago: Annales de Chimie, vol. xxxix. p. 369. *On the variable Intensity of Terrestrial Magnetism, and the influence of the Aurora Borealis upon it;* by Robert Were Fox. — Philosophical Transactions, 1831, p. 199.

([180]) p. 185. — "Brilliant and active coruscations of the Aurora Borealis," says Captain Back, "when seen through a hazy atmosphere, and exhibiting the prismatic colours, almost invariably affected the needle. On the contrary, a very bright Aurora, though attended by motion, and even tinged with a dullish red and a yellow in a clear blue sky, seldom produced any sensible change, beyond, at the most, a tremulous motion. A dense haze or fog, in conjunction with an active Aurora, seemed uniformly favourable to the disturbance of the needle, and a low temperature was favourable to brilliant and active coruscations. On no occasion during two winters was any sound heard to accompany the motions. The Aurora was frequently seen at twilight, and as often to the eastward as to the westward; clouds, also, were often perceived in the day-time, in form and disposition very much resembling the Aurora." — *Narrative of the Arctic Land Expedition.*

([181]) p. 187. — Faraday; *On the Diamagnetic character of Flame and Gases.*

([182]) p. 187. — "The Aurora Borealis is certainly in some measure a magnetical phenomenon; and if iron were the only substance capable of exhibiting magnetic effects, it would follow that some ferruginous particles must exist in the upper regions of the atmosphere. The light usually attending this magnetical meteor may possibly be derived from electricity, which may be the immediate cause of a change in the distribution of the magnetic fluid, contained in the ferruginous vapours which are imagined to float in the air." — Lecture on Magnetism: Young's *Lectures on Nat. Philosophy*, p. 533.

([183]) p. 188. — *On the supposed influence of Magnetism and Chemical Action:* by Robert Hunt. — Philosophical Magazine, vol. xxxii. No. 215. 1848.

([184]) p. 188. — Those bodies which are attracted by a magnet, as iron is, are called *magnetic bodies.* Those which are, on the contrary, repelled by the same power, are termed *diamagnetic bodies.* On these Dr. Faraday remarks: "Of the substances which compose the crust of the earth, by far the greater portion belong to the diamagnetic class; and though ferruginous and other magnetic matters, being more energetic in their action, are more striking in their phenomena, we should be hasty in assuming that, therefore, they overrule entirely the effect of the former bodies. As regards the ocean, lakes, rivers, and the atmosphere, they will exert their peculiar effect almost uninfluenced by any magnetic matter in them, and as respects the rocks and mountains, their diamagnetic influence is perhaps greater than might be anticipated. I mentioned that, by adjusting water and a salt of iron together, I obtained a solution inactive in air; that is, by a due association of the forces of a body from each class, water and a salt of iron, the magnetic force of the latter was entirely counteracted by the diamagnetic force of the former, and the mixture was neither attracted nor repelled. To produce this effect, it required that more than 48.6 grains of crystallized protosulphate of iron should be added to ten cubic inches of water (for these proportions gave a solution which would set equatorially); a quantity so large, that I was greatly astonished on observing the power of the water to overcome it. It is not, therefore, at all unlikely that many of the masses which form the crust of this our globe, may have an excess of diamagnetic power, and act accordingly." — *On new magnetic actions, and on the magnetic condition of all matter:* by Michael Faraday, D.C.L., F.R.S., &c. — Philosophical Transactions, Jan. 1846, vol cxxxvii. p. 41.

([185]) p. 189. — Ibid.

([186]) p. 190. — *On the Diamagnetic conditions of Flame and Gases,* by Michael Faraday, F.R.S.,; and *On the motions presented by Flame when under Electro-Magnetic Influence,* by Professor Zantedeschi.— Philosophical Magazine, 1847, pp. 401–421.

([187]) p. 190. — *On Diamagnetism;* by Professor Plücker, of Bonn. — Philosophical Magazine, July, 1848.

([188]) p. 191. — A few examples taken from Dr. Faraday's paper will show this: —

Nitrogen being acted on was manifestly diamagnetic in relation to common air when both were of the same temperature. Oxygen appears to be magnetic in common air. Hydrogen proved to be clearly and even strongly diamagnetic. Its diamagnetic state shows, in a striking point of view, that gases, like solids, have peculiar and distinctive degrees of diamagnetic force. Carbonic acid gas is diamagnetic in air. Carbonic oxide was carefully freed from carbonic acid before it was used, and it appears to be more diamagnetic than carbonic acid. Nitrous oxide was moderately, but clearly diamagnetic in air. Olefiant gas was diamagnetic. The coal gas of London is very well diamagnetic, and gives exceedingly good and distinct results. Sulphurous acid gas is diamagnetic in air. Muriatic acid gas was decidedly diamagnetic in air. *On the Diamagnetic Conditions of Flame and Gases:* Philosophical Magazine, 1847, p. 409.

([189]) p. 193. — For illustration of this I must refer to my own Memoir, *Researches on the Influence of Magnetism and Voltaic Electricity on Crystallization, and other conditions of matter*, in the Memoirs of the Geological Survey of Great Britain, &c., vol. i.

([190]) p. 194. — In a work published by Mr. Evan Hopkins, entitled *On the connection of Geology with Terrestrial Magnetism*, will be found many valuable practical observations, made in this country and the gold and silver districts of America; but the views taken by the author are open to many objections.

([191]) p. 195. — See a Notice by Faraday of Morichini's Experiments in *Relations of Light to Magnetic Force.* — Philosophical Transactions, vol. cxxxvii. p. 15. See also Mr. Christie *On Magnetic influence in the Solar Rays.* — Philosophical Transactions, vol. cvii. p. 219; vol. cxix. p. 379.

([192]) p. 195. — Sir David Brewster *On Magnetism;* republished from the Encyclopædia Britannica.

([193]) p. 195. — The whole of the title of Kircher's book will convey some idea of the subjects embraced: Athanasii Kircheri Societatis Jesu Magnes, sive de Arte Magneticâ: opus tripartitum, quo Universa Magnetis Natura ejusque in omnibus Scientiis et Artibus usus novâ methodo explicatur: ac præterea e viribus et prodigiosis effectibus Magneticarum aliarumque abditarum Naturæ Motionum in Elementis, Lapidibus, Plantis, Animalibus elucescentium: multa hucusque incognita Naturæ Arcana, per Physica, Medica, Chymica, et Mathematica omnis generis Experimenta recluduntur. Editio Tertia: ab ipso Authore recognita emendataque, ac multis novorum Experimentorum Problematibus aucta. Romæ, 1654.

([194]) p. 195.—The following are the titles of the concluding chapters of Kircher's book: *De magnetismo solis et lunæ in maria. De magneticâ vi plantarum. De insitionis magneticis miraculis. De magnetismo virgulæ auriferæ seu divinatoriæ. De plantis heliotropiis eorumque magnetismo. De magnetismo rerum medicinalium. De vi attractivâ potentiæ imaginativæ. De magnetismo musicæ. De magnetismo amoris.*

([195]) p. 196.—"For these reasons it appears most natural to seek their origin in the sun, the source of all living activity, and our conjecture gains probability from the preceding remarks on the daily oscillations of the needle. Upon this principle, the sun may be conceived as possessing one or more magnetic axes, which, by distributing the force, occasion a magnetic difference in the earth, in the moon, and all those planets whose internal structure admits of such a difference. Yet, allowing all this, the main difficulty seems not to be overcome, but merely removed from the eyes to a greater distance; for the question may still be asked, with equal justice, Whence did the sun acquire its magnetic force? And if from the sun we have recourse to a central sun, and from that again to a general magnetic direction throughout the universe, having the Milky Way for its equator, we but lengthen an unrestricted chain, every link of which hangs on the preceding link, no one of them on a point of support. All things considered, the following mode of representing the subject appears to me most plausible. If a single globe were left to move alone freely in the immensity of space, the opposite forces existing in its material structure would soon arrive at an equilibrium conformable to their nature, if they were not so at first, and all activity would soon come to an end. But if we imagine another globe to be introduced, a mutual relation will arise between the two; and one of its results will be a reciprocal tendency to unite, which is designated and sometimes thought to be explained by the merely descriptive word attraction. Now would this tendency be the only consequence of this relation? Is it not more likely that the fundamental forces, being drawn from their state of indifference or rest, would exhibit their energy in all possible directions, giving rise to all kinds of contrary action? The electric force is excited, not by friction alone, but also by contact, and probably also, though in smaller degrees, by the mutual action of two bodies at a distance; for contact is nothing but the smallest possible distance, and that, moreover, only for a few small particles. Is it not conceivable that magnetic force may likewise originate in a similar manner? When the natural philosopher and the mathematician pay regard to no other effect of the reciprocal relation between two bodies at a distance, except the tendency to unite, they proceed logically, if their investigations require nothing more than moving power; but should it be maintained that no other energy can be developed between two such bodies, the assertion will need proof, and the proof will be hard to find." — The above is a translation from Hansteen's work *On Magnetism.*

(196) p. 198. — See article *Animal Magnetism*, Encyclopædia Britannica, and Mr. Braid's papers *On Hypnotism*, published in the Medical Times.

(197) p. 200. — All the phenomena connected with volcanic action, and the theories connected therewith, will be found in Dr. Daubeny's work, *A description of active and extinct Volcanoes, of Earthquakes, and of Thermal Springs*, 1848.

(198) p. 201. — Graham's *Elements of Chemistry.* New Edition.

(199) p. 202. — Ibid.; and Brande's *Manual.*

(200) p. 203. — Of these *tables of attraction* the following may be taken as a specimen: —

SULPHURIC ACID.

Baryta.	Lime.
Strontia.	Magnesia.
Potassa.	Ammonia.
Soda.	

It thus appears that baryta separates sulphuric acid from its compounds with all inferior substances, and that ammonia is separated from the acid by all that are above it.

(201) p. 204. — Berthollet; *Essai de Statique Chimique.* 1803. Sir Humphrey Davy, in his *Elements of Chemical Philosophy*, has given an excellent review of the views of Berthollet.

(202) p. 208. — *On certain combinations of a new acid, formed of Azote, Sulphur, and Oxygen*: by J. Pelouze. Translated from Annales de Chimie, vol. xvi., for Scientific Memoirs, vol. i. p. 470. *Some ideas of a new force acting in the combinations of Organic Compounds*, by Berzelius: Annales de Chimie, vol. lxi. The conclusion come to by this eminent chemist is expressed in the following translation: "This new power, hitherto unknown, is common both in organic and inorganic nature. I do not believe that it is a power which is entirely independent of the electro-chemical affinities of the sub-

stance. I believe, on the contrary, that it is merely a new form of it; but so long as we do not see their connection and mutual dependence, it will be more convenient to describe it by a separate name. I shall, therefore, call it *catalytic power:* I shall also call *catalysis* the decomposition of bodies by this force — in the same way as the decomposition of bodies by chemical affinity is termed analysis."

(203) p. 209. — Berzelius; *Annales de Chimie*, vol. lxi.

(204) p. 210. — *On Transformations produced by Catalytic bodies:* by Lyon Playfair, Esq.; Phil. Mag., vol. xxxi. p. 191. 1847. — "Facts have been brought forward to show that there is at least as much probability in the view that the catalytic force is merely a modified form of chemical affinity exerted under peculiar conditions, as there is in ascribing it to an unknown power, or to the communication of an intestine motion to the atoms of a complex molecule. Numerous cases have been cited, in which the action results, when the assisting or catalytic body is not in a state of change; and attempts have been made to prove, by new experiments, that the catalytic power exercises its peculiar power by acting in the same direction as the body decomposing, or entering into union, but under conditions in which its own affinity cannot always be gratified."

(205) p. 213. — Consult Graham's Chemistry, *On Combining Proportions.*

(206) p. 215. — *Memoir on Atomic Volume and Specific Gravity.* Messrs. Lyon Playfair and Joule. — Philosophical Magazine, vol. xxvii. p. 453, or Transactions of Chemical Society of London. *Observations* on the above, by Professor de Marignac. — Bibliothèque Universelle, Feb. 1846. *On the Relation of the Volume of bodies, in the solid state, to their equivalents, or atomic weights:* by Professor Otto. *Studies on the connection between the atomic weights, crystalline form, and density of bodies:* by M. Filhol. Translated for the Cavendish Society, and published in their Chemical Reports and Memoirs.

(207) p. 215. — *Comptes Rendus de l'Académie des Sciences*, 1840, No. 5. A good translation of Dumas's Memoir appeared in the Philosophical Magazine, from which I extract the following familiar exposition of the laws of substitution: "Let me make a comparison, drawn from a familiar order of ideas. Let us put ourselves in the place of a man overlooking a game at chess without the slightest knowledge of the game. He would soon remark that the pieces must be used according to positive rules. In chemistry, the equivalents are our pieces, and the law of substitutions one of the rules which preside over their moves. And as, in the oblique move of the pawns, one pawn must be substituted for another, so in the

phenomena of substitution, one element must take the place of another. But this does not hinder the pawn from advancing without taking anything, as the law of substitution does not hinder an element from acting on a body without displacing or taking the place of any other element that it may contain." — *Memoir on the Law of Substitutions, and Theory of Chemical Types.*

([208]) p. 217. — Liebig's *Chemistry, in its application to Agriculture and Physiology*: translated by Lyon Playfair, Ph. D. *Animal Chemistry, or Chemistry in its application to Physiology and Pathology*: by Justus Liebig; translated by William Gregory.

([209]) p. 223. — Dr. Priestley appears to have been the first to observe the peculiar property of the diffusion of gases. Dr. Dalton, however, first drew attention to the important bearing of this fact on natural phenomena, and he published his views on *The Miscibility of Gases*, in the Manchester Memoirs, vol. v. The following extract from his memoir *On the Constitution of the Atmosphere*, will exhibit its bearings: —

It may be worth while to contrast this view of the constitution of the atmosphere with the only other one, as far as I know, that has been entertained.

According to one view,

1. The volumes of each gas found at the surface of the earth are proportional to the whole weights of the respective atmospheres.

Azote	79
Oxygen	21
Aqueous vapour	1.33
Carbonic Acid	.10
	101.43

According to one view,

2. The altitude of each atmosphere differs from that of every other, and the proportions of each in the compound atmosphere gradually vary in the ascent.

3. When two atmospheres are mixed, they take their places according to their specific gravity, not in separate strata, but intermixedly. There is, however, a separate stratum of the specifically lighter atmosphere a the summit over the other.

According to the other view,

1. The volume of each gas found at the surface of the earth, *multiplied by its specific gravity*, is proportional to the whole weight of the respective atmospheres.

Azote	76.6
Oxygen	23.4
Aqueous vapour	0.83
Carbonic Acid	0.15
	100.88

According to the other view,

2. The altitude of each atmosphere is the same, and the proportion of each in the compound atmosphere is the same at all elevations.

3. When two atmospheres are mixed, they continue so without the heavier manifesting any disposition to separate and descend from the lighter.

(210) p. 223.—The discussion of this question, commenced by Arago in his *Eloge*, was continued by Lord Brougham in his *Lives of Watt and Cavendish*, and by Mr. Vernon Harcourt, in his address as President of the British Association, and more recently in his *Letter to Lord Brougham.* Watt's *Letters* on the subject have been since published under the superintendence of Mr. Muirhead.

(211) p. 224.—See several papers *On Ozone*, by Professor Schönbein, in the Philosophical Magazine, and in the Reports of the British Association. Consult a paper by the Author: *Athenæum*, September, 1849.

(212) p. 227.—Iodide of silver has been found at Albarradon, near Mazapil, in Mexico. Iodide of mercury, of a fine lemon-yellow colour, has been discovered in the sandstone of Casas, Viegas, Mexico. Algers; Phillips' *Mineralogy.*

(213) p. 228.—Stahl, taking up the obscure notions of Becher and Van Helmont, supposed the phenomena of combustion to be due to phlogiston. He imagined that, by combination with phlogiston, a body was rendered combustible, and that its disengagement occasioned combustion, and after its evolution there remained either an acid or an earth: thus sulphur was, by this theory, supposed to be composed of phlogiston and sulphuric acid, and lead of the calx of lead and phlogiston, &c.

(214) p. 229.—Being called upon by the Solicitor for the Admiralty to examine into the causes of the fire which destroyed the *Imogene* and *Talavera*, in Devonport Arsenal, I discovered a bin under the roofing which covered these ships, in which there had been accumulating for a long period all the refuse of the wheelwrights' and painters' shop; and it was quite evident that spontaneous combustion had taken place in the mass of oiled oakum, sawdust, anti-attrition, and old sail-cloth there allowed to accumulate.

(215) p. 230.—*Researches on Flame:* Sir H. Davy's collected works.

(216) p. 232.—See note, *ante, On the Chemical Theory of Respiration.*

(217) p. 236.—At the request of the British Association, a committee has undertaken the investigation of this subject. Experiments are now being carried on by Dr. Daubeny, in the Botanic Gardens at Oxford, and by the Author, at his residence, Stockwell. Dr. Daubeny, in his report made at the meeting of the British Association at Birmingham, appears disposed to consider ten per cent. of carbonic acid in excess as destructive to the growth of ferns.

(218) p. 236.—See memoir *On the Pilchard*, by Mr. Couch, in the Reports of the Royal Cornwall Polytechnic Society.

(219) p. 239. — "This scale, in which the humidity of the air is expressed, is the simple natural scale in which air at its maximum of humidity (*i. e.*, when it is saturated with vapour) is reckoned as = 100, and air absolutely deprived of moisture as = 0; the intermediate degrees are given by the fraction 100 × actual tension of vapour ÷ tension required for the saturation of the air at its existing temperature. Thus, if the air at any temperature whatsoever contains vapour of half the tension, which it would contain if saturated, the degree is 50; if three-fourths, then 75; and so forth. Air of a higher temperature is capable of containing a greater quantity of vapour than air of less temperature; but it is the proportion of what it does contain, to what it would contain if saturated, which constitutes the measure of its dryness or humidity. The capacity of the air to contain moisture being determined by its temperature, it was to be expected that an intimate connection and dependence would be found to exist between the annual and diurnal variations of the vapour and of the temperature." — Sabine, *On the Meteorology of Toronto:* Reports of British Association, vol. xiii. p. 47. *The Temperature Tables:* by Prof. W. H. Dove; Reports for 1847 should be consulted.

(220) p. 240. — Sir David Brewster's *Optics* and memoirs in the Philosophical Transactions. Sir John Herschel's treatise on *Light*, Encyclopædia Metropolitana.

(221) p. 241. — *On the colour of steam under certain circumstances;* by Professor Forbes; Philosophical Magazine, vol. xiv. p. 121, vol. xv. p. 25. In the first paper the following remarks occur: "I cannot doubt that the colour of watery vapour under certain circumstances is the principal or only cause of the red colour observed in clouds. The very fact that that colour chiefly appears in the presence of clouds is a sufficient refutation of the only explanation of the phenomena of sunset and sunrise having the least plausibility given by the optical writers. If the red light of the horizontal sky were simply complementary to the blue of a pure atmosphere, the sun ought to set red in the clearest weather, and then most of all; but experience shows that a lurid sunrise or sunset is *always* accompanied by clouds or diffused vapours, and in a great majority of cases occurs when the changing state of previously transparent and colourless vapour may be inferred from the succeeding rain. In like manner terrestrial lights, seen at a distance, grow red and dim when the atmosphere is filled with vapour soon to be precipitated. Analogy, applied to the preceding observations, would certainly conduct to a solution of such appearances; for I have

remarked that the existence of vapour of high tension is by no means essential to the production of colour, though of course a proportionally greater thickness of the medium must be employed to produce a similar effect when the elasticity is small."

([222]) p. 243. — *On the Law of Diffusion of Gases:* by Thomas Graham, M.A., F.R.S., &c.; Edinburgh Philosophical Transactions, 1832. *Sur l'Action Capillaire des Fissures, &c.:* by Döbereiner; Annales de Chimie, xxiv. 332.

([223]) p. 245. — *Electro-chemical Researches on the Decompositions of the Earths, with observations on the Metals obtained from the Alkaline Earths, and on the Amalgam procured from Ammonia:* by Sir Humphrey Davy; Philosophical Transactions, 1808, and collected works, vol. v. p. 102.

([224]) p. 247. — *Elements of Chemical Philosophy:* by Sir H. Davy.

([225]) p. 252. — *Preliminary Discourse;* Sir J. F. W. Herschel. Lardner's Cabinet Cyclopædia.

([226]) p. 254. — *Geological Researches:* by Sir Henry De la Beche, C.B. (*Degradation of Mountains,* p. 167.) *Geological Manual,* p. 184. *Principles of Geology;* by Sir Charles Lyell, 7th Edition, p. 150, 686. *On the Denudation of South Wales, and the adjacent countries of England:* by Professor Andrew Ramsay; Memoirs of the Geological Survey and Museum of Practical Geology, vol. i. p. 297.

([227]) p. 255. — Fownes, *On the existence of Phosphoric Acid in Rocks of Igneous origin;* Phil. Trans. 1844. p. 53. Nesbitt, *Quarterly Journal of the Chemical Society.*

([228]) p. 256. — *On the Vegetation of the Carboniferous period as compared with that of the present day; On some peculiarities in the structure of Stigmaria: Remarks on the Structure and Affinities of some Lepidostrobi:* by Dr. Hooker; Memoirs of the Geological Survey, &c., vol. ii. pp. 387, 431, 440.

([229]) p. 259. — See Owen, Quarterly Journal of the Geological Society, No. 6, p. 96. Dr. Buckland, Geological Transactions, vol. iii. p. 220. *The Wonders of Geology:* by Dr. Mantell, vol. ii. p. 493.

([230]) p. 259. — *Report on British Fossil Mammalia:* by Richard Owen, Esq., F.R.S.; British Association Reports, vols. xi. xii.

([231]) p. 260. — *Notice on the Iguanodon, a newly discovered fossil reptile from the Sandstone of Tilgate Forest, in Sussex:* by Gideon Mantell, Esq., F.R.S., &c.; Philosophical Transactions, vol. cxv. p. 179. *On the Structure of Teeth, &c.:* by Professor Owen.

([232]) p. 260. — Dr. Mantell, *Wonders of Geology. Geology of the South-east of England.*

([233]) p. 261. — *Geological Researches; Geological Manual:* by Sir Henry Thomas De la Beche, C.B., &c.

([234]) p. 261. — Ibid.

([235]) p. 262. — *Experimental Researches on the Production of Silicon from Paracyanogen:* by Samuel Brown, M.D.; Transactions of the Royal Society of Edinburgh, vol. xv. p. 229. *Experiments on the alleged conversion of Carbon into Silicon:* by R. H. Brett, Ph. D., and J. Denham Smith, Esq.; Philosophical Magazine, vol. xix. p. 295, New Series. See also Dr. Brown's reply to the above, ibid., p. 388.

([236]) p. 262. — *Geology; Introductory, Descriptive, and Practical:* by Prof. Ansted, vol. ii. p. 22.

([237]) p. 263. — *The Wonders of Geology:* by Dr. Mantell, vol. i. p. 162. *Bridgewater Treatise:* by Dr. Buckland. Dr. J. J. Kemp, and Dr. A. V. Klipstein, *On the Dionotherium;* Darmstadt, 1836. Cuvier and De Blainville have also carefully described the fossil remains of this animal.

([238]) p. 266. — See Professor Ramsay's memoir *On Denudation,* quoted Note 220.

([239]) p. 266. — "The distances to which river water, more or less charged with detritus, would flow over sea-water, will depend upon a variety of obvious circumstances. Captain Sabine found discoloured water, supposed to be that of the Amazons, three hundred miles distant in the ocean from the embouchure of that river. It was about 126 feet deep. Its specific gravity was = 1.0204, and the specific gravity of the sea-water = 1.0262. This appears to be the greatest distance from land at which river-water has been detected on the surface of the ocean. If rivers, containing mechanically suspended detritus, flowed over sea-water in lines which, in general terms, might be called straight, the deposit of transported matter which they carried out would also be in straight lines. If, however, they be turned aside by an ocean current, as was the case with that observed by Captain Sabine, the detritus would be thrown, and cover an area corresponding in a great degree with the sweep which the river has been compelled to make out of the course, that its impulse, when discharged from its embouchure, might lead it to take. Supposing the velocity with which this river-water was moving has been correctly estimated at about three miles per hour, it is not a little curious to consider that the agitation and resistance of its particles should be sufficient to keep finely comminuted solid matter mechanically suspended, so that it would not be disposed freely to part with it, except at its junction with the sea-water over which it flows, and where, from friction, it is sufficently retarded. So that a river, if it can preserve a given amount of velocity flowing over the sea, may deposit no very large amount of mechanically suspended detritus in its course from the embouchure, where it is ultimately stopped. Still, however, though the deposit may not be so abundant as at first sight would appear probable, the constant accumulation of matter, however inconsiderable at any given time, must produce an appreciable effect during the lapse of ages." — Sir Henry De la Beche's Geological Researches, p. 72.

([240]) p. 267. — Sir J. F. W. Herschel; *Preliminary Treatise.*

([241]) p. 270. — *Fauna Antiqua Sivalensis. Being the Fossil Zoology of the Sewalik Hills in the North of India:* by Hugh Falconer and Proby T. Cautley. 1844.

([242]) p. 273. — Percy Bysshe Shelley.

([243]) p. 275. — *Experiments on the production of dephlogisticated air from water with various substances:* by Lieut.-General Sir Benjamin, Count of Rumford; Phil. Trans., vol. lxxvii. p. 84.

([244]) p. 275. — *Experiments upon Vegetables, discovering their great power of purifying the common air in the Sunshine, and of injuring it in the Shade and at Night; to which is joined, A new method of examining the accurate degrees of Salubrity of the Atmosphere,* by John Ingenhousz, Councillor of the Court, and Body Physician to their Imperial and Royal Majesties, F.R.S., &c. London: printed for P. Elmsley, in the Strand, and H. Payne, Pall Mall, 1779.

([245]) p. 275. — *The Kingdoms of Nature, their life and affinity:* by Dr. C. G. Carus; Scientific Memoirs, vol. i. p. 223.

([246]) p. 275. — In *Biologie,* by G. R. Treviranus, vol. ii. p. 302, the following passage occurs: "If we expose spring water to the sun in open or even closed transparent vessels, after a few days bubbles rise from the bottom, or from the sides of the vessel, and a green crust is formed at the same time. Upon observing this crust through a microscope, we discover a mass of green particles, generally of a round or oval form, very minute, and overlaid with a transparent mucous covering, some of them moving freely, whilst others, perfectly similar to these, remain motionless and attached to the sides of the vessel. This motion is sometimes greater than at others. The animalcules frequently lie as if torpid, but soon recover their former activity."

([247]) p. 276. — *On the Structure of the Vegetable Cell:* by Mohl. — Scientific Memoirs, vol. iv. p. 113. *Outlines of Structural and Physical Botany:* by Henfrey.

[248] p. 276. — Dr. Carus, in the memoir already quoted, says: "But since, in the organization of the earth, light and air, as constituting a second integrant part, stand opposed to gravitation, and since the plant bears a relation, not only to gravitation, but to light also, when its formation is complete it will necessarily present a second anatomical system, namely, that of the spiral vessels, which have been very justly considered, of late, as the organs that perform in plants the functions of nerves."

[249] p. 277.—Mr. Crosse's Experiments in the Journal of the London Electrical Society, and Mr. Weeke's in the Electrical Magazine, and a communication appended to *Explanations: a Sequel to the Vestiges of the Natural History of Creation.*

[250] p. 282. — *Die Metamorphose der Pflanzen:* Goethe, sect. 78.

[251] p. 283. — Lindley's *Elements of Botany.*

[252] p. 283. — See the very curious experiments of C. Matteucci. Traduit et extrait du "*Cimento.*"— Archives des Sciences Physiques et Naturelles; *Quelques Expériences sur la Respiration des Plantes.* Nov. 1846.

[253] p. 285. — Consult *Rural Economy*, by J. B. Boussingault; *The Chemical and Physiological Balance of Organic Nature*, by Dumas and Boussingault; and *Agricultural Chemistry*, by Liebig.

[254] p. 287. — See note, *ante*, 122.

[255] p. 293. — *Reports of the Fauna of the Ægean:* by Professor Forbes. — Reports of the British Association. *On the Physical Conditions affecting the distribution of Life in the Sea and the Atmosphere, &c.:* by Dr. Williams. Swansea.

[256] p. 295. — *The Vestiges of the Natural History of Creation.*

[257] p. 295. — *General Outline of the Animal Kingdom:* by Professor Thomas Rymer Jones, F.Z.S.

[258] p. 301. — In addition to the memoirs already referred to, Note 249, see Carlisle, *On the battery of the Torpedo, governed by a voluntary muscle.* — Phil. Trans., vol. xcv. p. 11. Todd, *Experiments on the Torpedo of the Cape of Good Hope.* —Ibid., vol. cvi. p. 120. Todd, *Experiments on the Torpedo Electricus at La Rochelle.* — Ibid., vol. cvii. p. 32.

[259] p. 301. — For a concise account of these experiments see *Elements of Natural Philosophy:* by Golding Bird, A.M., M.D., &c., 3rd Edition, chap. xx. p. 336. In this work all the most recent researches are given, and the authorities referred to; see also Matteucci's interesting papers, already quoted.

(260) p. 301. — *On the laws according to which the mixing of fluids, and their penetration into permeable substances, occurs, with special reference to the processes in the Human and Animal Organism*, by Julius Vogel, of Giessen: translated for the Cavendish Society. Liebig, *On the Motion of the Juices in the Animal body.*

(261) p. 308. — *A general Outline of the Animal Kingdom:* by Thomas Rymer Jones, p. 54, et seq.

(262) p. 311. — "As to the polishing and grooving of hard rocks, it has lately been ascertained that glaciers give rise to these effects when pushing forward sand, pebbles, and rocky fragments, and causing them to grate along the bottom. Nor can there be any doubt that icebergs, when they run aground on the floor of the ocean, imprint similar marks upon it."—*Principles of Geology, or the modern changes of the Earth and its Inhabitants considered as illustrative of Geology:* by Charles Lyell, M.A., F.R.S. *Travels through the Alps of Savoy, and other parts of the Pennine Chain, with Observations on the Phenomena of Glaciers:* by James D. Forbes, F.R.S.

(263) p. 314. — This may be readily proved by the following simple but instructive experiment: Take two pairs of watch-glasses; into one pair put a solution of nitrate of silver, into the other, a weak solution of iodide of potassium; connect the silver solution of each pair with a potash one by a film of cotton, and carry a platina wire from one glass into the other. Place one series in sunshine, and the other in a dark place. After a few hours, it will be found that the little galvanic arrangement in the dark will exhibit, around the platina wire, a very pretty crystallization of metallic silver, but no such change is observable in the other exposed to light. If a yellow glass is interposed between the glass and the sunshine, the action proceeds as when in the dark. This experiment is naturally suggestive of many others, and it involves some most important considerations.

(264) p. 314. — In cases of violent death, it is often found the gastric juice has, in a few hours, dissolved portions of the stomach. — Dr. Budd's Lecture before the College of Physicians.

(265) p. 314. — Faraday's *Experimental Researches*, vol. i.; from which a quotation has already been made, showing the enormous quantity of electricity which is latent in matter.

(266) p. 314. — *On the Motion of Gases:* by Professor Graham, F.R.S. — Phil. Trans., vol. cxxxvii. p. 573.

INDEX.

AT A MEETING OF THE

BRITISH ASSOCIATION FOR THE ADVANCEMENT OF SCIENCE.

Mr. Lyell in the Chair.

"Mr. Murchison gave an account of the investigations and discoveries of Mr. Hugh Miller of Cromarty (now Editor of the "Witness") in the Old Red Sandstone. Various members of a great family of fishes, existing only in a deposit of the very highest antiquity, had been discovered by Mr. Miller, Dr. Fleming, Dr. Malcolmson, and other gentlemen. M. Agassiz had found these fishes to be characterized by the peculiarity of not having the vertebral column terminated at the centre of the tail, as in the existing species, but at its extremity. He spoke in the highest terms of Mr. Miller's perseverance and ingenuity as a geologist. With no other advantage than a common education, by a careful use of his means, he had been able to give himself an excellent education, and to elevate himself to a position which any man in any sphere of life might well envy. Mr. Murchison added, that he had seen some of Mr. Miller's papers on Geology, written in a style so beautiful and poetical, as to throw plain geologists like himself into the shade. (Cheers.) The fish discovered by Mr. Miller, one or two fine specimens of which were on the table, was yet without a name; and perhaps M. Agassiz, who would now favor them with a description of the class to which it belonged, would assign it one.

"M. Agassiz stated, that since he first saw, five or six years ago, the fishes of the old deposits, they had increased to such an extent as to enable him to connect them with one large geological epoch. This had been still further established by their having been found in the same formation by Mr. Murchison in Russia, and Mr. Miller in Ross-shire. These fishes were characterized in the most curious way he had ever seen. After briefly adverting to their peculiarities, as illustrated by the specimens on the table, M. Agassiz proposed to call Mr. Miller's the *Pterichthys Milleri*. In the course of a subsequent conversation, the learned Professor added, that in lately examining the eggs of the salmon, he had observed that in the fœtal state of these fishes they have that unequally divided condition of tail which characterizes so large a portion of the fishes in the older strata, and which becomes so rare in the fishes of the cretaceous and post-cretaceous formations.

"Dr. Buckland said, he had never been so much astonished in his life by the powers of any man as he had been by the geological descriptions of Mr. Miller, which had been shown to him in the "Witness" newspaper by his friend Sir C. Menteath. That wonderful man described these objects with a facility which made him ashamed of the comparative meagreness and poverty of his own description in the "Bridgewater Treatise," which had cost him hours and days of labor. He (Dr. Buckland) would give his left hand to possess such powers of description as this man; and if it pleased Providence to spare his useful life, he, if any one, would certainly render the science attractive and popular, and do equal service to Theology and Geology. It must be gratifying to Mr. Miller to hear that his discovery had been assigned his own name by such an eminent authority as M. Agassiz; and it added another proof of the value of the meeting of the Association, that it had contributed to bring such a man into notice." — *Extract from the Report of the Proceedings of the Association.*

GOULD, KENDALL & LINCOLN, Publishers, Boston.

PROSPECTUS.

The editors are so situated as to have access to all the scientific publications of America, Great Britain, France, and Germany; and have also received, for the present volume, the approbation as well as the counsel and personal contributions of many of the ablest scientific men in this country, among whom are PROFESSORS AGASSIZ, HORSFORD, and WYMAN, of Harvard University, and they have the promise in future, from many scientific gentlemen, of articles not previously published elsewhere. They have not confined themselves to an examination of Scientific Journals and Reports, but have drawn from every source which furnished any thing of scientific interest. For those who have occasion for still further researches, they have furnished a copious Index to the scientific articles in the American and European Journals, and, moreover, they have prepared a list of all books pertaining to Science which have appeared originally, or by republication, in the United States, during the year. A classified List of Patents, and brief obituaries of men distinguished in Science or Art, who have recently died, render the work still more complete. They have also taken great pains to make the General Index to the whole as full and correct as possible.

It will thus be seen, that the plan of the "ANNUAL OF SCIENTIFIC DISCOVERY" is well designed to make it what it purports to be, a *substantial summary of the discoveries in Science and Art;* and no pains have been spared on the part of the editors to fulfil the design, and render it worthy of patronage.

As the work is not intended for scientific men exclusively, but to meet the wants of the general reader, it has been the aim of the editors that the articles should be brief and intelligible to all; and to give authenticity, the source from whence the information is derived is generally stated. Although they have used all diligence to render this first issue as complete as possible, in its design and execution, yet they hope that experience, and the promised aid and co-operation from the many gentlemen interested in its success, will enable them in future to improve both on the plan and the details.

☞ *The work will hereafter be published annually on the first of March, and will form a handsome duodecimo volume of about* 360 *pages, with an engraved likeness of some distinguished man of science.* Price $1.00, *paper, or in substantial cloth binding,* $1.25.

On the receipt of $1.00 *the publishers will forward a copy* in paper covers, *by mail,* post paid.

GOULD, KENDALL & LINCOLN, *Publishers,*

59 WASHINGTON STREET, BOSTON.

ANNUAL OF SCIENTIFIC DISCOVERY.

NOTICES OF THE PRESS.

"Nothing which has transpired in the scientific world during the past year, seems to have escaped the attention of the industrious editors. We do not hesitate to pronounce the work a highly valuable one to the man of Science."—*Boston Journal.*

"This is a highly valuable work. We have here brought together in a volume of moderate size, all the leading discoveries and inventions which have distinguished the past year. Like the hand on the dial-plate, 'it marks the progress of the age.' The plan has our warmest wishes for its eminent success."—*Christian Times.*

"A most acceptable volume."—*Transcript.*

"The work will prove of unusual interest and value."—*Traveller.*

"We have in our possession the ledger of progress for 1849, exhibiting to us in a condensed form, the operations of the world in some of the highest business transactions. To say that its execution has been worthy of its aim is praise sufficient."—*Springfield Republican.*

"To the artist, the artisan, the man of letters, it is indispensable, and the general reader will find in its pages much valuable material which he may look for elsewhere in vain." —*Boston Herald.*

"We commend it as a standard book of reference and general information, by those who are so fortunate as to possess it."—*Saturday Rambler.*

"A body of useful knowledge, indispensable to every man who desires to keep up with the progress of modern discovery and invention."—*Boston Courier.*

"Must be a most acceptable volume to every one, and greatly facilitate the diffusion of useful knowledge."—*Zion's Herald.*

"A most valuable and interesting popular work of science and art."—*Washington National Intelligencer.*

"A rich collection of facts, and one which will be eagerly read. The amount of information contained within its pages is very large."—*Evening Gazette.*

"Such a key to the progress and facts of scientific discovery will be everywhere welcomed."—*New York Commercial Advertiser.*

"A most valuable, complete, and comprehensive summary of the existing facts of science; it is replete with interest, and ought to have a place in every well appointed library."—*Worcester Spy.*

"We commend it to all who wish what has just been found out; to all who would like to discover something themselves, and would be glad to know how: and to all who think they have invented something, and are desirous to know whether any one else has been before hand with them."—*Puritan Recorder.*

"This is one of the most valuable works which the press has brought forth during the present year. A greater amount of useful and valuable information cannot be obtained from any book of the same size within our knowledge."—*Washington Union.*

"This important volume will prove one of the most acceptable to our community that has appeared for a long time."—*Providence Journal.*

"This is a neat volume and a useful one. Such a book has long been wanted in America. It should receive a wide-spread patronage."—*Scientific American, New York.*

"It meets a want long felt, both among men of science and the people. No one who feels any interest in the intellectual progress of the age, no mechanic or artisan, who aspires to excel in his vocation, can afford to be without it. A very copious and accurate index gives one all needed aid in his inquiries."—*Phil. Christian Chronicle.*

"One of the most useful books of the day. Every page of it contains some useful information, and there will be no waste of time in its study."—*Norfolk Democrat.*

"It is precisely such a work as will be hailed with pleasure by the multitude of intelli gent readers who desire to have, at the close of each year, a properly digested record of its progress in useful knowledge. The project of the editors is an excellent one, and deserves and will command success."—*North American, Philadelphia.*

"Truly a most valuable volume."—*Charleston (S. C.) Courier.*

"There are few works of the season whose appearance we have noticed with more sincere satisfaction than this admirable manual. The exceeding interest of the subjects to which it is devoted, as well as the remarkably thorough, patient and judicious manner in which they are handled by its skilful editors, entitle it to a warm reception by all the friends of solid and useful learning."—*New York Tribune.*

GOULD, KENDALL & LINCOLN, PUBLISHERS, BOSTON.

SECOND EDITION, REVISED.

THE EARTH AND MAN:

LECTURES ON COMPARATIVE PHYSICAL GEOGRAPHY.

BY ARNOLD GUYOT.

TESTIMONIALS

IN FAVOR OF THE FIRST EDITION OF THIS WORK.

From Prof. Louis Agassiz, of Harvard University.

"GENTLEMEN: — I understand that you are publishing the Lectures of Prof. Guyot on Physical Geography. Having been his friend from childhood, as a fellow-student in college, and as a colleague in the same university, I may be permitted to express my high sense of the value of his attainments. Mr. Guyot has not only been in the best school, that of Ritter and Humboldt, and become familiar with the present state of the science of our earth, but he has himself, in many instances, drawn new conclusions from the facts now ascertained, and presented most of them in a new point of view. Several of the most brilliant generalizations developed in his lectures are his; and will not only render the study of geography more attractive, but actually show it in its true light, namely, as the science of the relations which exist between nature and man, throughout history; of the contrasts observed between the different parts of the globe; of the laws of horizontal and vertical forms of the dry land, in its contact with the sea; of climate, &c. It would be highly serviceable, it seems to me, for the benefit of schools and teachers, that you should induce Mr. Guyot to write a series of graduated text-books of geography, from the first elements up to a scientific treatise. It would give new life to these studies in this country, and be the best preparation for sound statistical investigations."

From Prof. George Ticknor, Boston.

"I was very glad to learn, that you intend to publish Mr. Guyot's Lectures on Physical Geography. Their familiar and simple manner will, I hope, cause them to be used in our schools, where I think their modest learning and religious philosophy will make them an excellent foundation for the study of all geography, as it is now taught, and especially of that higher geography which connects itself with the destinies of the whole human race."

From George S. Hillard, Esq., Boston.

"Professor Guyot's Lectures are marked by learning, ability and taste. Familiar with the labors of all who have gone before him, he has been himself an extensive and accurate observer. His bold and comprehensive generalizations rest upon a careful foundation of facts. The essential value of his statements is enhanced by his luminous arrangement, and by a vein of philosophical reflection which gives life and dignity to dry details. Such a work as his Lectures will furnish will be a valuable accession to our literature. I cannot think so lightly of the judgment and taste of our community, as to entertain any doubt of its success. To teachers of youth it will be especially important. They may learn from it how to make Geography, which I recall as the least interesting of studies, one of the most attractive; and I earnestly commend it to their careful consideration."

TESTIMONIALS.

From Prof. C. C. Felton, of Harvard University.

"I cannot help believing that by publishing the volume you will render an acceptable service to an intelligent and appreciating public. The original lectures, in point of style, are characterized by simplicity and elegance."

From Charles Sumner, Esq., Boston.

"It was my good fortune to hear several of these Lectures, as delivered, and I have since read them all in print. The instruction and satisfaction which they have afforded to me, I shall be glad to see within the reach of others. Beyond the intrinsic interest of the subject, they have the charm of simplicity and clearness, while the elevated sentiment which inspires the lecturer, and which naturally belongs to his theme, makes science seem like a Christian preacher. Most truly do I thank him for teaching so persuasively the duties of the superior races of men towards the races which are inferior in the scale of creation—to succor, protect, and elevate, not to subdue, depress, and enslave. Thus has he drawn from these founts of science the divine lesson of charity and good-will to men."

From Prof. Benjamin Peirce, of Harvard University.

"Having heard or read the greater portion of Professor Guyot's Lectures on Physical Geography, I cannot forbear expressing the strong feeling which I have of their scientific and literary merits, and of the importance of their publication. He has set himself to work at the foundation of an almost new science, with the ability and simplicity of a true master; he has developed profound and original views, with the most enlarged variety and richness of illustration, and in the most attractive and eloquent forms of language. His ingenious investigations, sustained by faithful and conscientious research, are an invaluable addition to science; while the vivid and picturesque earnestness of their utterance cannot fail to charm the least learned of his readers."

From Rev. Edward N. Kirk, Boston.

"Many will hail with delight the introduction of Prof. Guyot to the great field of education in our country. His Lectures on Physical Geography will open a new career of study to many of our teachers, as well as learners; and will form to them a true scientific basis for the study of History. And if Mr. Guyot can follow this work by some *elementary books for schools*, he will increase his claims to the gratitude of the country which is now ready to adopt him."

From George B. Emerson, Esq., Boston.

"I received, some time ago, a copy of Prof. Guyot's excellent work on Physical Geography, which the business of my school prevented me from acknowledging. I avail myself of my earliest leisure to thank you for it. The work contains much which has not been made accessible to English readers, and much of original generalization, which render it a most valuable work. It ought to be in the hands of every teacher of Geography. It will enable him to read and understand the high lessons which the study of nature is calculated to teach, but which, without some guiding philosophical principles, are apt to be missed, or to be lost sight of. It will enable him, in very many particulars, to give an interest to the study of Geography, which mere barren, unrelated, unassociated facts can never possess to the youthful student. It brings the imagination, and the desire to search into causes, to the aid of the memory.

"Much of the chapters relating to the distribution of rains is, so far as I know, now for the first time laid before the American reader by the American press. The publication of the work will mark an era in the teaching of Geography."

CHAMBERS'S

CYCLOPÆDIA OF ENGLISH LITERATURE.

A SELECTION OF THE CHOICEST PRODUCTIONS OF ENGLISH AUTHORS, FROM THE EARLIEST TO THE PRESENT TIME: CONNECTED BY A CRITICAL AND BIOGRAPHICAL HISTORY.

EDITED BY ROBERT CHAMBERS.

ASSISTED BY ROBERT CARRUTHERS AND OTHER EMINENT GENTLEMEN.

Complete in two imperial octavo volumes, of more than fourteen hundred pages of double column letterpress, and upwards of three hundred elegant illustrations.

This valuable work has now become so generally known and appreciated, that there need scarcely be any thing said in commendation, except to those who have not yet seen it.

The work embraces about One Thousand Authors, chronologically arranged and classed as Poets, Historians, Dramatists, Philosophers, Metaphysicians, Divines, etc., with choice selections from their writings, connected by a Biographical, Historical, and Critical Narrative; thus presenting a complete view of English Literature, from the earliest to the present time. Let the reader open where he will, he cannot fail to find matter for profit and delight, which, for the most part, too, repeated perusals will only serve to make him enjoy the more. We have indeed infinite riches in a little room. No one, who has a taste for literature, should allow himself, for a trifling consideration, to be without a work which throws so much light upon the progress of the English language. The selections are gems — a mass of valuable information in a condensed and elegant form.

EXTRACTS FROM COMMENDATORY NOTICES.

From W. H. Prescott, Author of "Ferdinand and Isabella." "The plan of the work is very judicious. * * It will put the reader in the proper point of view, for surveying the whole ground over which he is travelling. * * Such readers cannot fail to profit largely by the labors of the critic who has the talent and taste to separate what is really beautiful and worthy of their study from what is superfluous."

"I concur in the foregoing opinion of Mr. Prescott." — *Edward Everett.*

"It will be a useful and popular work, indispensable to the library of a student of English literature." — *Francis Wayland.*

"We hail with peculiar pleasure the appearance of this work, and more especially its republication in this country at a price which places it within the reach of a great number of readers." — *North American Review.*

"This is the most valuable and magnificent contribution to a sound popular literature that this century has brought forth. It fills a place which was before a blank. Without it, English literature, to almost *all* of our countrymen, educated or uneducated, is an imperfect, broken, disjointed mass. Much that is beautiful — the most perfect and graceful portions, undoubtedly — was already possessed; but it was not a whole. Every intelligent man, every inquiring mind, every scholar, felt that the foundation was missing. Chambers's Cyclopædia supplies this radical defect. It begins with the beginning; and, step by step, gives to every one who has the intellect or taste to enjoy it a view of English literature in all its complete, beautiful, and perfect proportions." — *Onondaga Democrat, N. Y.*

"We hope that teachers will avail themselves of an early opportunity to obtain a work so well calculated to impart useful knowledge, with the pleasures and ornaments of the English classics. The work will undoubtedly find a place in our district and other public libraries; yet it should be the 'vade mecum' of every scholar." — *Teachers' Advocate, Syracuse, N. Y.*

"The work is finely conceived to meet a popular want, is full of literary instruction, and is variously embellished with engravings illustrative of English antiquities, history, and biography. The typography throughout is beautiful." — *Christian Reflector, Boston.*

"The design has been well executed by the selection and concentration of some of the best productions of English intellect, from the earliest Anglo-Saxon writers down to those of the present day. No one can give a glance at the work without being struck with its beauty and cheapness." — *Boston Courier.*

"We should be glad if any thing we can say would favor this design. The elegance of the execution feasts the eye with beauty, and the whole is suited to refine and elevate the taste. And we might ask, who can fail to go back to its beginning, and trace his mother-tongue from its rude infancy to its present maturity, elegance, and richness?" *Christian Mirror, Portland.*

*** The Publishers of the AMERICAN Edition of this valuable work desire to state that, besides the numerous pictorial illustrations in the English Edition, they have greatly enriched the work by the addition of fine steel and mezzotint engravings of the heads of Shakspeare, Addison, Byron; a full length portrait of Dr. Johnson, and a beautiful scenic representation of Oliver Goldsmith and Dr. Johnson. These important and elegant additions, together with superior paper and binding, must give this a decided preference over all other editions.

GOULD, KENDALL, & LINCOLN, PUBLISHERS, BOSTON.

The Works of John Harris, D.D.

THE PRE-ADAMITE EARTH: Contributions to Theological Science. Price 85 cents.

"It is a book for thinking men. It opens new trains of thought to the reader — puts him in a new position to survey the wonders of God's works; and compels Natural Science to bear her decided testimony in support of Divine Truth." — *Phila. Ch. Observer.*

MAN PRIMEVAL; Or, the Constitution and Primitive Condition of the Human Being. A Contribution to Theological Science. With a finely engraved portrait of the author; 12mo. cloth, price $1.25.

*** This is the second volume of a series of works on Theological Science. The first was received with much favor — the present is a continuation of the principles which were seen holding their way through the successive kingdoms of primeval nature, and are here resumed and exhibited in their next higher application to individual man.

"His copious and beautiful illustrations of the successive laws of the Divine Manifestation, have yielded us inexpressible delight." — *London Eclectic Review.*

THE GREAT COMMISSION; Or, the Christian Church constituted and charged to convey the Gospel to the World. A Prize Essay. With an Introductory Essay, by W. R. WILLIAMS, D.D. Price $1.00

"Of the several productions of Dr. Harris, — all of them of great value, — that now before us is destined, probably, to exert the most powerful influence in forming the religious and missionary character of the coming generations. But the vast fund of argument and instruction comprised in these pages will excite the admiration and inspire the gratitude of thousands in our own land as well as in Europe. Every clergyman and pious and reflecting layman ought to possess the volume, and make it familiar by repeated perusal." *Boston Recorder.*

"His plan is original and comprehensive. In filling it up, the author has interwoven facts with rich and glowing illustrations, and with trains of thought that are sometimes almost resistless in their appeals to the conscience. The work is not more distinguished for its arguments and its genius, than for the spirit of deep and fervent piety that pervades it." *The Day-Spring.*

THE GREAT TEACHER; Or, Characteristics of our Lord's Ministry. With an Introductory Essay, by H. HUMPHREY, D.D. Tenth thousand. Price 85 cents.

"The book itself must have cost much meditation, much communion on the bosom of Jesus, and much prayer. Its style is, like the country which gave it birth, beautiful, varied, finished, and everywhere delightful. But the style of this work is its smallest excellence. It will be read: it ought to be read. It will find its way to many parlors, and add to the comforts of many a happy fireside. The reader will rise from each chapter, not able, perhaps, to carry with him many striking remarks or apparent paradoxes, but he will have a sweet impression made upon his soul, like that which soft and touching music makes when every thing about it is appropriate. The writer pours forth a clear and beautiful light, like that of the evening light-house, when it sheds its rays upon the sleeping waters, and covers them with a surface of gold. We can have no sympathy with a heart which yields not to impressions delicate and holy, which the perusal of this work will naturally make." *Hampshire Gazette.*

MISCELLANIES; Consisting principally of Sermons and Essays. With an Introductory Essay and Notes, by J. BELCHER, D.D. Price 75 cents.

"Some of these essays are among the finest in the language; and the warmth and energy of religious feeling manifested in several of them, will render them peculiarly the treasure of the closet and the Christian fireside." — *Bangor Gazette.*

MAMMON; Or, Covetousness, the Sin of the Christian Church. A Prize Essay. Price 45 cents. Twentieth thousand.

*** This masterly work has already engaged the attention of churches and individuals, and receives the highest commendations.

ZEBULON; Or the Moral Claims of Seamen stated and enforced. Edited by Rev. W. M. ROGERS and D. M. LORD. Price 25 cents.

*** A well written and spirit-stirring appeal to Christians in favor of this numerous, useful, and long neglected class.

THE ACTIVE CHRISTIAN; Containing the "Witnessing Church," "Christian Excellence," and "Means of Usefulness," three popular productions of this talented author. Price 31 cents.

Ripley's Notes,---Cruden's Concordance.

THE FOUR GOSPELS, WITH NOTES. Chiefly Explanatory; intended principally for Sabbath School Teachers and Bible Classes, and as an aid to Family Instruction. By HENRY J. RIPLEY, Newton Theol. Institution. Seventh Edition. Price $1.25.

*** This work should be in the hands of every student of the Bible, especially every Sabbath School and Bible Class teacher. It is prepared with special reference to this class of persons, and contains a mass of just the kind of information wanted.

"The undersigned, having examined Professor Ripley's Notes on the Gospels, can recommend them with confidence to all who need such helps in the study of the sacred Scriptures. Those passages which all can understand are left 'without note or comment,' and the principal labor is devoted to the explanation of such parts as need to be explained and rescued from the perversions of errorists, both the ignorant and the learned. The practical suggestions at the close of each chapter, are not the least valuable portion of the work. Most cordially, for the sake of truth and righteousness, do we wish for these Notes a wide circulation.

BARON STOW,	R. H. NEALE,	R. TURNBULL,
DANIEL SHARP,	J. W. PARKER,	N. COLVER.
WM. HAGUE,	R. W. CUSHMAN,	

THE ACTS OF THE APOSTLES, WITH NOTES. Chiefly Explanatory. Designed for Teachers in Sabbath Schools and Bible Classes, and as an Aid to Family Instruction. By Prof. HENRY J. RIPLEY. Price 75 cents.

"The external appearance of this book, — the binding and the printed page, — 'it is a pleasant thing for the eyes to behold.' On examining the contents, we are favorably impressed, first, by the wonderful perspicuity, simplicity, and comprehensiveness of the author's style; secondly, by the completeness and systematic arrangement of the work, in all its parts, the 'remarks' on each paragraph being carefully separated from the exposition; thirdly, by the correct theology, solid instruction, and consistent explanations of difficult passages. The work cannot fail to be received with favor. These Notes are much more full than the Notes on the Gospels, by the same author. A beautiful map accompanies them." — *Christian Reflector, Boston.*

CRUDEN'S CONDENSED CONCORDANCE. A Complete Concordance to the Holy Scriptures; by ALEXANDER CRUDEN, M.A. A New and Condensed Edition, with an Introduction; by Rev. DAVID KING, LL.D. Fifth Thousand. Price in Boards, $1.25; Sheep, $1.50.

*** "This edition is printed from English plates, and is a full and fair copy of all that is valuable in Cruden as a Concordance. The principal variation from the larger book consists in the exclusion of the Bible Dictionary, which has long been an incumbrance, and the accuracy and value of which have been depreciated by works of later date, containing recent discoveries, facts, and opinions, unknown to Cruden. The condensation of the quotations of Scripture, arranged under their most obvious heads, while it diminishes the bulk of the work, greatly facilitates the finding of any required passage.

"Those who have been acquainted with the various works of this kind now in use, well know that Cruden's Concordance far excels all others. Yet we have in this edition of Cruden, the best made better. That is, the present is better adapted to the purposes of a Concordance, by the erasure of superfluous references, the omission of unnecessary explanations, and the contraction of quotations, &c.; it is better as a manual, and is better adapted by its price to the means of many who need and ought to possess such a work, than the former larger and expensive edition." — *Boston Recorder.*

"The new, condensed, and cheap work prepared from the voluminous and costly one of Cruden, opportunely fills a chasm in our Biblical literature. The work has been examined critically by several ministers, and others, and pronounced complete and accurate."
Baptist Record, Phila.

"This is the very work of which we have long felt the need. We obtained a copy of the English edition some months since, and wished some one would publish it; and we are much pleased that its enterprising publishers can now furnish the student of the Bible with a work which he so much needs at so cheap a rate." — *Advent Herald, Boston.*

"We cannot see but it is, in all points, as valuable a book of reference, for ministers and Bible students, as the larger edition." — *Christian Reflector, Boston.*

"The present edition, in being relieved of some things which contributed to render all former ones unnecessarily cumbrous, without adding to the substantial value of the work, becomes an exceedingly cheap book." — *Albany Argus.*

Valuable School Books.

BLAKE'S FIRST BOOK IN ASTRONOMY. Designed for the Use of Common Schools. By J. L. BLAKE, D.D. Illustrated by Steel Plate Engravings. 8vo. cloth back. Price 50 cents.

From E. Hinckley, Professor of Mathematics in Maryland University.

"I am much indebted to you for a copy of the First Book in Astronomy. It is a work of utility and merit, far superior to any other which I have seen. The author has selected his topics with great judgment, — arranged them in admirable order, — exhibited them in a style and manner at once tasteful and philosophical. Nothing seems wanting, — nothing redundant. It is truly a very beautiful and attractive book, calculated to afford both pleasure and profit to all who may enjoy the advantage of perusing it."

From B. Field, Principal of the Hancock School, Boston.

"I know of no other work on Astronomy so well calculated to interest and instruct young learners in this sublime science."

From James F. Gould, A.M., Principal of the High School for Young Ladies, Baltimore, Md.

"I shall introduce your First Book in Astronomy into my Academy in September, consider it decidedly superior to any elementary work of the kind I have ever seen."

From Isaac Foster, Instructor of Youth, Portland.

"I have examined Blake's First Book in Astronomy, and am much pleased with it. A very happy selection of topics is presented in a manner which cannot fail to interest the learner, while the questions will assist him materially in fixing in the memory what ought to be retained. It leaves the most intricate parts of the subject for those who are able to master them, and brings before the young pupil only what can be made intelligible and interesting to him."

"The illustrations, both pictorial and verbal, are admirably intelligible; and the definitions are such as to be easily comprehended by juvenile scholars. The author has interwoven with his scientific instructions much interesting historical information, and contrived to dress his philosophy in a garb truly attractive. — *N. Y. Daily Evening Journal.*

"We are free to say, that it is, in our opinion, decidedly the best work we have any knowledge of, on the sublime and interesting subject of Astronomy. The engravings are executed in a superior style, and the mechanical appearance of the book is extremely prepossessing. The knowledge imparted is in language at once chaste, elegant, and simple — adapted to the comprehension of those for whom it was designed. The subject matter is selected with great judgment, and evinces uncommon industry and research. We earnestly hope that parents and teachers will examine and judge for themselves, as we feel confident they will coincide with us in opinion. We only hope the circulation of the work will be commensurate with its merits." — *Boston Evening Gazette.*

"The book now before us contains forty-two short lessons, with a few additional ones. which are appended in the form of problems, with a design to exercise the young learner in finding out the latitude and longitude on the terrestrial globe. We do not hesitate to recommend it to the notice of the superintending committees, teachers, and pupils of our public schools. The definitions in the first part of the volume are given in brief and clear language, adapted to the understanding of beginners."—*State Herald, Portsmouth, N. H.*

BLAKE'S NATURAL PHILOSOPHY. Being Conversations on Philosophy, with the addition of Explanatory Notes, Questions for Examination, and a Dictionary of Philosophical Terms. With twenty-eight steel Engravings. By J. L. BLAKE, D.D. 12mo. sheep. Price 67 cents.

*** Perhaps no work has contributed so much as this to excite a fondness for the study of Natural Philosophy in youthful minds. The familiar comparisons, with which it abounds, awaken interest, and rivet the attention of the pupil.

From Rev. J. Adams, President of Charleston College, S. C.

"I have been highly gratified with the perusal of your edition of Conversations on Natural Philosophy. The Questions, Notes, and Explanations of Terms, are valuable additions to the work, and make this edition superior to any other with which I am acquainted. I shall recommend it wherever I have an opportunity."

"We avail ourselves of the opportunity furnished us by the publication of a new edition of this deservedly popular work, to recommend it, not only to those instructors who may not already have adopted it, but also generally to all readers who are desirous of obtaining information on the subjects on which it treats. By Questions arranged at the bottom of the pages, in which the collateral facts are arranged, he directs the attention of the learner to the principal topics. Mr. Blake has also added many Notes, which illustrate the passages to which they are appended, and the Dictionary of Philosophical Terms is a useful addition." — *U. S. Literary Gazette.*

Valuable School Books.

THE YOUNG LADIES' CLASS BOOK. A Selection of Lessons for Reading in Prose and Verse. By E. BAILEY, A.M., late Principal of the Young Ladies' High School, Boston. Stereotyped Edition. 12mo. sheep. Price 83⅓ cents.

From the Principals of the Public Schools for Females, Boston.

"GENTLEMEN: — We have examined the Young Ladies' Class Book with interest and pleasure; with interest, because we have felt the want of a Reading Book expressly designed for the use of females; and with pleasure, because we have found it well adapted to supply the deficiency. In the selections for a Reader designed for boys, the eloquence of the bar, the pulpit, and the forum may be laid under heavy contribution; but such selections, we conceive, are out of place in a book designed for females. We have been pleased, therefore, to observe, that in the Young Ladies' Class Book such pieces are rare. The high-toned morality, the freedom from sectarianism, the taste, richness, and *adaptation* of the selections, added to the neatness of its external appearance, must commend it to all; while the practical teacher will not fail to observe that diversity of style, together with those peculiar *points*, the want of which, few, who have not felt, know how to supply.

Respectfully yours, BARNUM FIELD, ABRAHAM ANDREWS, R. G. PARKER, CHARLES FOX"

From the Principal of the Mount Vernon School, Boston.

"I have examined with much interest the Young Ladies' Class Book, by Mr. Bailey and have been very highly pleased with its contents. It is my intention to introduce it into my own school; as I regard it as not only remarkably well fitted to answer its particular object as a book of exercises in the art of elocution, but as calculated to have an influence upon the character and conduct, which will be in every respect favorable.

JACOB ABBOTT."

"We were never so struck with the importance of having reading books for female schools, adapted particularly to that express purpose, as while looking over the pages of this selection. The eminent success of the compiler in teaching this branch, to which we can personally bear testimony, is sufficient evidence of the character of the work, considered as a selection of lessons in elocution; they are, in general, admirably adapted to cultivate the amiable and gentle traits of the female character, as well as to elevate and improve the mind." — *Annals of Education.*

"The reading books prepared for academic use, are often unsuitable for females. We are glad, therefore, to perceive that an attempt has been made to supply the deficiency; and we believe that the task has been faithfully and successfully accomplished. The selections are judicious and chaste; and so far as they have any moral bearing, appear to be unexceptionable." — *Education Reporter.*

ROMAN ANTIQUITIES AND ANCIENT MYTHOLOGY. By C. K. DILLAWAY, A.M., late Principal in the Boston Latin School. With Engravings. Eighth Ed., improved. 12mo. half mor. Price 67 cts.

From E. Bailey, Principal of the Young Ladies' High School, Boston.

"Having used *Dillaway's Roman Antiquities and Ancient Mythology* in my school for several years, I commend it to teachers with great confidence, as a valuable text-book on those interesting branches of education. E. BAILEY.'

"The want of a cheap volume, embracing a succinct account of ancient customs, together with a view of classical mythology, has long been felt. To the student of a language, some knowledge of the manners, habits, and religious feelings of the people whose language is studied is indispensably requisite. This knowledge is seldom to be obtained without tedious research or laborious investigation. Mr. Dillaway's book seems to have been prepared with special reference to the wants of those who are just entering upon a classical career; and we deem it but a simple act of justice to say, that it supplies the want, which, as we have before said, has long been felt. In a small duodecimo, of about one hundred and fifty pages, he concentrates the most valuable and interesting particulars relating to Roman antiquity; together with as full an account of heathen mythology as is generally needed in our highest seminaries. A peculiar merit of this compilation, and one which will gain it admission into our highly respectable *female* seminaries, is the total absence of all allusion, even the most remote, to the disgusting *obscenities* of ancient mythology; while, at the same time, nothing is omitted which a pure mind would feel interested to know. We recommend the book as a valuable addition to the treatises in our schools and academies." — *Education Reporter, Boston.*

"We well remember, in the days of our pupilage, how unpopular as a study was the volume of Roman Antiquities introduced in the academic course. It wearied on account of its prolixity, filling a thick octavo, and was the prescribed task each afternoon for a long three months. It was reserved for one of our Boston instructors to apply the condensing apparatus to this mass of erudition, and so to *modernize* the *antiquities* of the old Romans, as to make a befitting abridgment for schools of the first order. Mr. Dillaway has presented such a compilation as must be interesting to lads, and become popular as a text-book. Historical facts are stated with great simplicity and clearness; the most important points are seized upon, while trifling peculiarities are passed unnoticed."—*Am. Traveller.*

www.ingramcontent.com/pod-product-compliance
Lightning Source LLC
LaVergne TN
LVHW020107110826
845151LV00001B/85